I0466857

Energy and Conflicts: From Black Gold to the Electric Revolution

The Impact of Oil and

Electric Vehicles in Our World

Francisco José Hurtado Mayén

ISBN: 9798329331509

Content

Introduction: The Energy Journey

Energy has been and continues to be the engine that drives our modern societies. Since the dawn of civilization, the search for energy sources has defined the course of human history. Today, the intersection between energy and geopolitics is more evident than ever, influencing armed conflicts, political decisions, and the future direction of technological innovation. This book, "Energy and Conflict: From Black Gold to the Electric Revolution," aims to explore these complex and crucial dynamics.

Oil, known as the "black gold," has been a dominant force in shaping global politics for more than a century. Its vast reserves, located in a few strategic regions of the world, have been both a blessing and a curse to the countries that possess them. The struggle for control of these resources has triggered countless conflicts, causing human suffering and environmental devastation. From the Gulf War to contemporary tensions in the Middle East, the history of oil is replete with episodes of violence and rivalry.

As the world moves forward, we face an inescapable reality: the urgent need for a shift to more sustainable energy sources. This is where the electric vehicle comes into the picture as a potential solution to reduce our dependence on fossil fuels. Electric vehicles promise a cleaner, less oil-dependent and more environmentally friendly future. However, its widespread adoption is fraught with challenges: insufficient infrastructure, high upfront costs, and questions about battery sustainability.

This book not only charts the journey of oil and its global consequences, but also discusses the emerging role of electric vehicles in the energy transition. We'll explore how technological innovations, led by pioneers like Toyota, are changing the mobility landscape. Toyota, with its commitment to electrification, exemplifies the automotive industry's efforts to adapt to the demands of the 21st century.

Through a multidisciplinary approach, this book examines the geopolitics of oil, associated armed conflicts, and the environmental impact of fossil fuels. It also delves into the history, benefits, and obstacles of the electric vehicle, providing a comprehensive view of the current and future energy landscape.

By the end of this journey, we hope that readers will have a deeper understanding of the forces shaping our energy dependence and the possible solutions for a more sustainable and peaceful future. This analysis is not only relevant to academics and practitioners in the sector, but also to anyone interested in the future of our planet and the energy that powers it.

Welcome to this energy journey, where we will explore how the past and present are interconnected in the pursuit of a greener and safer future.

Context and relevance of the topic

In the 21st century, energy has become the central axis of global discussions on sustainability, security and economic progress. Energy sources, particularly fossil fuels such as oil, have been fundamental to the industrial and technological development of nations. However, the world's dependence on these non-renewable resources has created a number of economic, political and environmental challenges that threaten to undermine the gains made so far.

Global Oil Dependence

Since the discovery of oil in the second half of the nineteenth century, this resource has been the backbone of the world economy. Entire countries have built their economies around the exploitation, refining, and export of oil. Large oil reserves are concentrated in geopolitically volatile regions such as the Middle East, Venezuela and Russia, which has led to intense competition for control and access to these resources.

The Organization of the Petroleum Exporting Countries (OPEC), founded in 1960, has played a crucial role in regulating oil production and prices globally. OPEC's decisions directly affect the global economy, influencing the costs of production, transportation, and consumer goods around the world. Fluctuations in oil prices can trigger economic crises, affect inflation, and disrupt the trade balances of importing and exporting countries.

Armed Conflicts and the Geopolitics of Oil

Recent history is full of examples where oil has been a determining factor in armed conflicts. From the Gulf War in 1990-1991, through the invasion of Iraq in 2003, to the conflicts in Libya and Syria, oil has been both a cause and spoils of war. These conflicts not only result in significant human and economic losses, but also lead to massive population displacements and severe environmental damage.

The 2003 U.S.-led invasion of Iraq is a clear example of how oil interests can motivate military action. Despite official justifications for weapons of mass destruction, many analysts agree that control of Iraq's vast oil reserves was an important underlying factor. Such interventions not only destabilise the affected regions, but also have global repercussions, altering power dynamics and the distribution of energy resources.

Environmental Impact of Fossil Fuels

The extensive use of oil and other fossil fuels has had a devastating impact on the environment. The burning of these resources is the main source of greenhouse gas emissions, contributing significantly to global warming and climate change. In addition, oil spills and leaks in transport and storage infrastructure have caused ecological disasters that affect biodiversity and human health.

Dependence on fossil fuels also perpetuates air pollution, which is responsible for millions of premature deaths each year due to respiratory and cardiovascular diseases. Oil extraction, especially unconventional techniques such as hydraulic fracturing (fracking), involves additional risks, including groundwater contamination and the generation of induced earthquakes.

The Electric Vehicle Revolution

In response to these challenges, a global movement towards the adoption of cleaner and more sustainable energy has emerged. The electric vehicle has become the symbol of this energy transition. With significant advances in battery technology, energy efficiency, and cost reduction, electric vehicles offer a viable and greener alternative to traditional internal combustion vehicles.

Leading automotive companies, such as Toyota, have played a pivotal role in the promotion and development of electric and hybrid vehicles. The launch of the Toyota Prius in 1997 marked the beginning of a new era in the automotive industry, showing that it is possible to combine energy efficiency with performance and reliability. As technology advances, charging infrastructure expands, and costs decrease, electric vehicles are increasingly within reach of global consumers.

Relevance of the Topic

The relevance of this topic cannot be underestimated. The transition from an economy based on fossil fuels to one based on renewable energies and electric vehicles is essential to mitigate the effects of climate change and ensure a sustainable future for generations to come. In addition, this transition has the potential to reduce the incidence of armed conflict associated with competition for scarce energy resources.

Understanding the dynamics between oil, armed conflict, and the electric vehicle revolution provides a comprehensive view of the challenges and opportunities we face. This book aims not only to inform, but also to inspire readers to reflect on their role in this transition and how they can contribute to a greener and more peaceful future.

In short, the energy journey is a story of power, conflict, and innovation. As we move towards a future less dependent on fossil fuels, it is crucial to learn from the past and embrace the technologies and policies that allow us to build a more just and sustainable world.

Objectives of the book

The relationship between oil, armed conflict, and the electric vehicle revolution is a fascinating story that unfolds on multiple levels, from the quicksand of geopolitics to technological innovation labs. Oil has been, for more than a century, the engine that has driven world economies, but it has also been the spark that has ignited numerous conflicts. This vital but controversial resource has shaped foreign policy, sparked wars and left an indelible mark on our environment.

The journey begins with the exploration of how oil has been the central axis of world geopolitics. From the vast deserts of the Middle East to the depths of Siberia, oil fields have dictated the balance of power between nations. The Organization of the Petroleum Exporting Countries (OPEC) has played a crucial role in regulating oil production and prices, influencing the global economy in profound and often conflicting ways. Countries such as the United States, Russia, and OPEC members have seen their foreign policies and defense strategies shaped by the need to ensure a steady supply of this valuable resource.

Oil-related armed conflicts are no less important. The Gulf War, the invasion of Iraq, and the conflicts in Libya and Syria are just a few examples of how competition for oil resources has sparked tensions and wars. These wars have not only devastated entire regions, but have also taken an immense human and economic toll. Millions of lives have been lost or displaced, and vital infrastructure has been destroyed, leaving deep scars that will take generations to heal. Moreover, these conflicts have exacerbated global instability, creating a vicious cycle of violence and resource dependence.

The environmental impact of oil use is another grim chapter in this story. Greenhouse gas emissions and ecological disasters associated with oil spills and leaks have had devastating consequences for our planet. Oil pollution has poisoned oceans and rivers, affecting marine life and coastal communities. Emissions of CO_2 and other pollutants have contributed to climate change, warming the planet at an alarming rate and causing extreme weather events that threaten our ways of life.

But in the midst of this bleak panorama, a light of hope is emerging: the electric vehicle revolution. The history of electric vehicles begins with their first prototypes at the beginning of the twentieth century, but it is in recent decades that they have gained significant momentum. Advances in battery technology, the expansion of charging infrastructure, and incentive policies have made electric vehicles an increasingly viable and attractive alternative. The advantages are clear: they reduce emissions, are more efficient and, in the long term, can be more economical than internal combustion vehicles.

Among the pioneers in this revolution is Toyota. With the launch of the Prius, the first mass-produced hybrid vehicle, Toyota not only demonstrated the potential of electric vehicles, but also set a new standard in the automotive industry. The company has continued to invest in research and development, expanding its line of hybrid and electric vehicles and leading the way towards more sustainable mobility. Toyota's innovations have not only reduced emissions, but have also inspired other manufacturers to follow suit, accelerating the transition to a cleaner future.

As electric vehicles gain traction, global policies play a crucial role in their adoption. Governments around the world are implementing initiatives to promote the development and adoption of electric vehicles, from economic incentives to strict emissions regulations. These policies not only drive demand for electric vehicles, but also encourage innovation and investment in clean technologies.

The transition to a sustainable energy future also requires critical and active awareness. It's not just about embracing new technologies, it's about changing our mindsets and behaviors. Each of us can contribute to this transformation, from choosing more sustainable modes of transport to supporting policies that promote clean energy. By understanding the importance of the energy transition and taking steps to reduce our carbon footprint, we can contribute to a healthier and fairer future for all.

In conclusion, the relationship between oil, armed conflict, and the electric vehicle revolution is a complex but essential narrative. Through a multidisciplinary approach, we have explored how these elements interact and shape our world. From the geopolitics of oil to advances in sustainable mobility, we have seen how energy dynamics influence our society and our planet. This book aims to be a comprehensive and accessible guide, providing valuable tools for academics, professionals in the sector and anyone interested in the challenges and opportunities facing our society in the energy field. With determination, innovation and global collaboration, we can build a future where clean energy and sustainable mobility are not only possible, but the norm, creating a healthier, fairer and more prosperous planet for all.

Methodology and approach

At the heart of this book, "Energy and Conflict: From Black Gold to the Electric Revolution," is an intricate and fascinating narrative that seeks to unravel the deep interconnectedness between oil, armed conflict, and the electric vehicle revolution. Our goal is not only to provide detailed information, but also to foster a critical and nuanced understanding of how energetic dynamics influence our society and our planet.

To provide a comprehensive and balanced view, we have taken a multidisciplinary approach. We integrate perspectives from history, geopolitics, economics, science, technology and the environment. We look at how oil has shaped modern history and international relations, exploring armed conflicts, foreign policies, and power dynamics between nations. We assess the economic impact of oil and electric vehicles, considering markets, production, consumption and their global implications. We investigate technological advances in oil extraction and electric vehicle development, highlighting innovations in batteries and emission reduction technologies. And, of course, we examine the environmental effects of fossil fuel use and the potential benefits of electric vehicles in terms of pollution and sustainability.

The methodology used in this book is rigorous and exhaustive. We conduct a detailed literature review, using academic databases to ensure coverage of high-quality and relevant sources. Keywords included terms such as "oil and geopolitics," "armed conflicts over resources," "environmental impact of oil," "electric vehicles," and "energy transition." We selected sources published in the last ten years, in English and Spanish, to ensure that the data and analysis were current and relevant.

In addition to the literature review, we incorporate analysis of specific case studies to illustrate key concepts and provide concrete examples. These case studies range from armed conflicts such as the Gulf War and the invasion of Iraq, to innovations in electric vehicles, such as the development of the Toyota Prius. Interviews and testimonies from energy sector experts, economists, historians and automotive industry professionals provide a personal and in-depth perspective on the challenges and opportunities in the field of energy and sustainable mobility.

The collection and analysis of quantitative data was also critical. We collect information on oil production and consumption, greenhouse gas emissions, electric vehicle sales, and government policies from trusted sources such as the International Energy Agency (IEA), the U.S. Energy Information Administration (EIA), and the International Organization of Automobile Manufacturers (OICA).

For ease of reading and understanding, the book is structured into thematic chapters that follow a logical flow from the history and geopolitics of oil to the electric vehicle revolution and future prospects. Each chapter is designed to be self-contained, allowing readers to address topics sequentially or choose the chapters of greatest interest according to their preference.

In the first chapter, "Black Gold and Power", the importance of oil and its global distribution is introduced, highlighting the role of OPEC in the regulation of production and prices. The second chapter, "Armed Conflicts over Oil", explores the relationship between oil and armed conflicts, presenting historical and contemporary cases and analyzing the impact on the population and the environment.

The third chapter, "Wounds of the Earth: The Environmental Impact of Oil," investigates ecological spills and catastrophes, air pollution, and mitigation and recovery measures. "The Hour of Change: The Electric Vehicle on Stage" traces the history of the electric vehicle, its current advantages and challenges, as well as the infrastructure and supporting policies needed.

The chapter dedicated to Toyota, "Innovation and the Future: The Toyota Case," examines the company's strategies in electrification, its technological innovations, and the impact on the automotive industry. "Global Policies for a Sustainable Future" looks at the government initiatives, economic incentives, and emissions regulations that are shaping the EV market.

Finally, "Looking Forward: The Energy Revolution" addresses the role of renewable energies in the energy transition, alternatives to electric vehicles and the future of mobility and sustainable energy.

Together, these chapters offer a comprehensive and accessible guide to understanding the complex interrelationships between the geopolitics of oil, armed conflict, and the electric vehicle revolution. Through a multidisciplinary approach and fluid narrative, "Energy and Conflict: From Black Gold to the Electric Revolution" not only seeks to educate, but also inspire and empower readers to better understand the challenges and opportunities we face in the transition to a more sustainable energy future.

On this journey, we have learned that the transition to clean and sustainable energy is not only a matter of technology or economics, but also of social justice and equity. International cooperation, the development of adequate infrastructure, and the implementation of effective policies are essential to facilitate this transition. It is also crucial to continue research into emerging technologies and their societal impacts to ensure that all the benefits of the energy revolution are shared equitably.

In the end, the purpose of this book is to provide a comprehensive and detailed overview of the complex relationship between oil, armed conflict, and the electric vehicle revolution. In doing so, we hope to foster a critical and in-depth understanding of how energy dynamics influence our society and our planet, inspiring readers to actively participate in promoting a more sustainable and peaceful energy future.

Chapter 1: Black Gold and Power

What is oil and why is it so important?

Petroleum, commonly called "black gold," is a viscous, dark substance that has played a crucial role in the development of modern civilization. But what exactly is oil, and why has it been so vital to our progress?

Petroleum is a complex mixture of hydrocarbons that formed millions of years ago from the remains of prehistoric plants and animals. These organic remains were buried under layers of sediment and subjected to high pressures and temperatures over millions of years, slowly transforming into oil. This process, which has taken eons to complete, makes oil a non-renewable resource that cannot be replaced on a human scale.

The importance of oil lies in its versatility and high energy density. This means that a small amount of oil can generate a large amount of energy. Here are some of the reasons why oil is so crucial:

1. **Energy Source**: Oil is the main source of energy for the world. It is used to generate electricity, heat our homes, and provide power for various industries. Without oil, many of the conveniences and technologies we take for granted would not be possible.
2. **Transportation**: Gasoline and diesel, derived from petroleum, are the fuels that power most cars, trucks, planes, and ships.

The transportation industry is heavily dependent on oil, making it a vital resource for global mobility.

3. **Industry and Manufacturing**: Petroleum is the raw material for a wide range of products, from plastics and chemicals to medicines and fertilizers. Petrochemicals is a key industry that produces essential materials for countless everyday products.

4. **Global Economy**: Oil is an economic engine, influencing financial markets and international relations. Oil-rich countries can have a major influence on the global economy, and fluctuations in oil prices can affect global economic stability.

In short, oil has been fundamental to the economic and technological development of the modern world. However, their extraction and use have also generated significant challenges, including armed conflicts and environmental problems. As the world looks for more sustainable alternatives, the transition away from oil represents one of the greatest challenges of our time.

World oil reserves and production

Oil reserves are not evenly distributed on the planet. Some regions have vast amounts of oil while others have very little or nothing. This unequal distribution has had a profound impact on global geopolitics, influencing the economy and stability of nations.

Major Oil-Producing Regions

- **Middle East**: This region is home to some of the largest oil reserves in the world. Countries such as Saudi Arabia, Iran, Iraq and Kuwait are giants in oil production. Oil wealth has given these countries significant influence on the global economy and international political decisions.
- **North America**: The United States and Canada are major oil producers. In particular, the United States has experienced a boom in production due to hydraulic fracturing technology, known as fracking. This technological advance has made it possible to exploit previously inaccessible reserves, altering the balance of the world oil market.
- **Russia**: Russia is one of the largest producers and exporters of oil. Its vast reserves and production capacity allow it to play a crucial role in European and global energy policy. Oil and gas are strategic tools in Russian foreign policy.
- **Venezuela**: This South American country has huge oil reserves, although production has declined significantly in recent years

due to economic and political problems. Venezuela's vast oil reserves represent a paradox of wealth and misery, where an abundance of resources does not translate into economic prosperity.

Global Production and Consumption

Global oil production is huge, reaching approximately 94 million barrels per day in 2021. However, production does not always coincide with consumption. Some countries produce much more oil than they consume, while others rely heavily on imports. This discrepancy creates complex dynamics in international trade and energy security.

- **The United States**: It is the largest producer of oil, but also one of the largest consumers. Its high level of consumption is largely due to its car-based transportation system and highly developed industry. Energy self-sufficiency has been a political goal for decades, influenced by market fluctuations and international conflicts.
- **China**: China's rapid economic growth has fueled high demand for oil, making it one of the largest importers. Dependence on foreign oil is a strategic concern for China, which is looking to diversify its energy sources and increase its domestic production.
- **India**: Like China, India has seen a significant increase in oil consumption due to its economic growth and urbanization.

Demand for energy is on the rise, and dependence on imported oil presents economic and security challenges.

The relationship between oil production and consumption is crucial to understanding global market dynamics. Fluctuations in supply and demand can affect prices, economic stability, and countries' energy policies.

OPEC's role in the global market

The Organization of the Petroleum Exporting Countries (OPEC) is a key player in the global oil market. Founded in 1960, OPEC brings together some of the world's largest oil producers with the aim of coordinating and unifying their oil policies. Its influence on the regulation of oil production and prices has been significant, affecting economies around the world.

OPEC Objectives and Operation

- **Market Stability**: OPEC seeks to stabilize oil prices and avoid excessive fluctuations that can harm both producers and consumers. This stability is crucial for planning investments and maintaining economic growth.

- **Policy Coordination**: OPEC members coordinate their production levels to influence oil prices. This often involves reducing production when prices are low and increasing it when they are high. Through regular meetings and negotiations, OPEC tries to balance the global market.

- **Revenue Protection**: By keeping oil prices at levels that benefit its members, OPEC protects the revenues of producing countries, many of which rely heavily on oil exports for their economies. This protection is vital to finance development and social stability programs.

Impact on the Global Market

OPEC has significant power to influence the oil market. For example, during the 1973 oil crisis, OPEC drastically reduced production, leading to skyrocketing oil prices and having a profound impact on the global economy. This event highlighted the world's dependence on oil and vulnerability to OPEC's decisions.

However, OPEC cannot always control the market completely. The emergence of new producers and technologies, such as fracking in the United States, has challenged their dominance. In addition, internal differences among OPEC members sometimes make it difficult to make coordinated decisions. National interests can conflict with collective goals, creating tensions within the organization.

Challenges and Future of OPEC

The future of OPEC is full of challenges. The global transition to cleaner and more sustainable energy, growing oil production in non-member countries, and internal tensions among its members are factors that can affect its influence. Pressure to reduce carbon emissions and comply with international agreements on climate change is driving the search for alternatives to oil.

As the world moves toward a future with less dependence on oil, OPEC will have to adapt to a changing energy landscape. This could involve diversifying the economy in member countries, investing in renewable energy, and promoting cleaner technologies. However, as long as oil remains a crucial source of energy, OPEC will continue to be a major player on the global stage.

In short, oil has been and continues to be a powerful force in the global economy and geopolitics. Its influence extends from the formation of large fortunes to the generation of armed conflicts. Understanding the importance of oil, the dynamics of its production and consumption, and the role of OPEC is essential to understanding the world we live in and the challenges we face in the transition to a more sustainable energy future. This knowledge is not only vital for experts in the field, but also for anyone interested in the future of our planet and the complex interactions that shape our daily lives.

How Oil Has Sparked Wars

Oil, known as the "black gold," has not only been an economic engine, but also a significant source of conflict throughout modern history. Its strategic and economic value has led to numerous wars and international tensions. Competition for control of oil resources has been a constant in global geopolitics, triggering conflicts that have left deep traces on humanity.

The Strategic Value of Oil

Oil is essential not only to the functioning of the global economy, but also to the national security of many countries. The ability to maintain an adequate reserve of oil is crucial for military operations and for economic stability. This makes oil a resource of strategic interest, and its control becomes a priority for nations.

International Competition for Oil

Competition for oil can take many forms, from economic rivalry to direct military intervention. Countries seek to ensure access to this resource through strategic alliances, interventions in foreign conflicts, and in some cases, the direct invasion of oil-rich countries. This competition can destabilize entire regions and lead to protracted conflicts.

Geopolitics of Petroleum

The geographical location of oil reserves has largely determined global geopolitics. Regions with large oil reserves, such as the Middle East, have been focal points of international rivalries. World powers have sought to secure their influence in these regions through various strategies, including support for allied governments, military intervention, and the imposition of economic sanctions.

Historical cases: Gulf War, Iraq, Syria

To better understand how oil has triggered armed conflict, let's examine some of the most prominent cases in recent history.

Gulf War (1990-1991)

The Gulf War, also known as Operation Desert Storm, is a clear example of how oil can be a decisive factor in triggering an armed conflict.

Context of the Conflict

In August 1990, Iraq, under the leadership of Saddam Hussein, invaded and occupied Kuwait. This invasion was motivated in part by territorial and economic disputes, but control of Kuwait's vast oil supply was a critical factor. With the annexation of Kuwait, Iraq would have controlled a significant proportion of the world's oil reserves, which would alter the balance of power in the region.

International Response

The international community, led by the United States, responded quickly to the invasion. The main concern was that Saddam Hussein's control over Kuwait's and potentially Saudi Arabia's oil supplies could give him disproportionate influence over the global oil market. The international coalition formed to liberate Kuwait included many countries that depended on oil from the Persian Gulf.

Development of the Conflict

Operation Desert Storm began in January 1991, after negotiations and sanctions failed to persuade Iraq to withdraw from Kuwait. The war was swift and decisive, with coalition forces liberating Kuwait in just over a month. The military victory restored the status quo in the region, but it also left Iraq under severe sanctions and international surveillance.

Impact

The conflict highlighted the vulnerability of the world's oil supply and the willingness of world powers to use military force to protect their strategic interests in the region. The repercussions of the conflict included further militarization of the region and the installation of U.S. military bases in several Persian Gulf countries.

Invasion of Iraq (2003)

The invasion of Iraq in 2003 is another example of how oil can influence foreign policy and lead to armed conflict.

Context of the Conflict

In March 2003, a coalition led by the United States and the United Kingdom invaded Iraq, citing the existence of weapons of mass destruction and Iraq's links to terrorism. However, many analysts and critics argue that control of Iraqi oil was also a significant factor in the decision to invade.

Iraq's Oil Reserves

Iraq has some of the largest oil reserves in the world. Control of these reserves was strategic both for global energy interests and for the economic stability of the region. The prospect of a hostile regime controlling this vast wealth of resources was a significant concern for the Western powers.

Development of the Conflict

The invasion began in March 2003 and quickly toppled Saddam Hussein's regime. However, the occupation and reconstruction of Iraq proved to be much more complicated and costly than anticipated. The country was plunged into a protracted insurgency and sectarian violence, leaving tens of thousands dead and millions displaced.

Impact

The invasion of Iraq had profound repercussions. In terms of oil, Iraqi production was severely affected by war and insurgency, which contributed to volatility in global oil markets. In addition, the Iraq war generated a wide debate about the ethics of military intervention for economic and strategic reasons.

Conflict in Syria (2011–present)

The conflict in Syria, which began in 2011, is also linked in part to the control of energy resources in the region.

Context of the Conflict

The conflict in Syria began as a series of peaceful protests against the government of Bashar al-Assad, but quickly morphed into a civil war. The fighting has involved multiple actors, including rebel groups, the Syrian government, ISIS, and foreign powers such as the United States, Russia, and Iran.

Oil and the Syrian Conflict

Syria is not a large oil producer compared to its neighbors, but its strategic location and energy infrastructure are important. During the conflict, various factions, including ISIS, have sought to control oil facilities to fund their operations. Syrian oil has become a source of income for the actors in the conflict, prolonging the violence and complicating the resolution of the conflict.

International Intervention

Foreign powers have intervened in Syria for a variety of reasons, including fighting terrorism and protecting their own strategic interests in the region. Russia's intervention, in particular, has been motivated in part by a desire to maintain its influence in the Middle East and protect its energy interests.

Impact

The conflict in Syria has resulted in a humanitarian catastrophe, with hundreds of thousands dead and millions displaced. The struggle for control of energy resources has aggravated the situation, hampering peace and reconstruction efforts.

Human and social impact of these conflicts

Armed conflicts over oil not only have geopolitical and economic repercussions, but also inflict a devastating human and social cost. Wars motivated by the control of energy resources leave behind a trail of destruction that affects millions of people.

Displacement of Populations

One of the most immediate and visible effects of these conflicts is the massive displacement of populations. People are forced to leave their homes due to violence and insecurity. In Iraq, the invasion and subsequent insurgency led to the displacement of millions of Iraqis. In Syria, the civil war has created one of the largest refugee crises in recent history, with millions of people seeking asylum in neighbouring countries and Europe.

Human Losses and Suffering

Oil wars result in a tragic loss of human life. Conflicts in the Gulf, Iraq and Syria have left hundreds of thousands dead and wounded. In addition to direct casualties, wars cause suffering through the destruction of essential infrastructure such as hospitals, schools, and water and sanitation systems. Lack of access to basic services aggravates living conditions and increases mortality.

Destruction of Infrastructure

Critical infrastructure, including oil facilities, roads, bridges, and buildings, is often targeted in these conflicts. The destruction of this infrastructure not only disrupts the production and supply of oil, but also negatively affects the economy and the well-being of the civilian population. Rebuilding devastated infrastructure is a long and expensive process.

Economic and Social Impact

War and instability have devastating effects on the economies of the affected countries. The disruption of oil production and other economic sectors can lead to a prolonged recession. Unemployment is rising and economic opportunities are shrinking, which in turn can fuel more conflict and violence. Poverty and lack of resources exacerbate social tensions and can lead to cycles of violence and destabilization.

Psychological trauma

Psychological trauma is another significant impact of armed conflict. People who have lived in war zones often suffer from post-traumatic stress disorder (PTSD), anxiety, and depression. Children, in particular, are vulnerable to the psychological effects of war. Trauma can affect future generations and make it difficult for affected societies to recover and rebuild.

Humanitarian Crises

Oil conflicts generate humanitarian crises that require international responses. Organizations such as the United Nations, the Red Cross and numerous NGOs work to provide aid and support to affected populations. However, insecure conditions and the destruction of infrastructure often hamper humanitarian efforts.

Reconstruction and Reconciliation

Recovery from armed conflict is a long and complex process. Rebuilding infrastructure, the economy and society requires enormous resources and coordinated efforts. In addition, reconciliation and peacebuilding are essential to avoid future conflicts. This means addressing the underlying causes of conflict, fostering dialogue and cooperation, and supporting communities in building a sustainable and peaceful future.

Conclusion

Armed conflicts over oil are a reminder of the deep interconnections between natural resources, geopolitics, and human well-being. The struggle for control of oil has triggered devastating wars and left a lasting mark on affected societies. Understanding these conflicts is crucial to finding solutions that promote peace, stability and a more sustainable use of energy resources. As the world moves towards cleaner and more sustainable energy sources, it is essential to learn from these historic episodes and work to prevent them from happening again in the future.

Chapter 3: Wounds of the Earth: The Environmental Impact of Oil

Spills and ecological catastrophes

Oil, despite being a vital resource for the economy and technological progress, has had devastating effects on the environment. One of the most serious and visible aspects of this impact is oil spills, which result in ecological catastrophes of major proportions. These events not only destroy entire ecosystems, but also affect wildlife, human health, and local economies. Below, we explore in depth the causes, consequences, and significant examples of oil spills.

Causes of Oil Spills

Oil spills can occur for a variety of reasons, many of which are related to the extraction, transportation, and storage of this resource.

1. Oil rig accidents

Oil rigs, both onshore and offshore, are common spill scenarios. These accidents can be caused by mechanical failures, human error, adverse weather conditions, or problems in the design and maintenance of infrastructure.

2. Oil Tanker Accidents

Transporting oil by sea is a high-risk activity. Oil tankers can suffer accidents due to collisions, groundings, storms, ship structure failures, or fires on board. These accidents can release large amounts of oil into the sea, causing extensive damage.

3. Pipelines and Leaks

Pipelines that transport oil from extraction areas to refineries and distribution centers can rupture or leak. These can be caused by corrosion, seismic movements, damage from human activities (such as construction) or sabotage.

4. Storage Failures

Storing oil in tanks can also be a source of spills. Tanks can leak due to structural failure, corrosion, or human error. Tanks in areas prone to natural disasters, such as earthquakes or floods, are especially at risk.

Consequences of Oil Spills

Oil spills have devastating and far-reaching consequences for the environment, wildlife, human health, and local economies.

1. Impact on Marine Ecosystems

Oil spills at sea are especially damaging to marine ecosystems. Oil can form a layer on the surface of the water, blocking sunlight and affecting the photosynthesis of aquatic plants. Corals, fish, mollusks, and other marine organisms can be severely damaged or killed due to exposure to oil and its toxic components.

2. Wildlife Effects

Marine wildlife, such as birds, marine mammals, and fish, are extremely vulnerable to oil spills. Birds can become covered in oil, which affects their ability to fly, float, and maintain body temperature. Marine mammals can ingest oil or inhale toxic fumes, which can cause serious health problems or death.

3. Water and Soil Pollution

Oil spills on land can contaminate soil and freshwater sources, affecting terrestrial flora and fauna. Toxic compounds in petroleum can infiltrate the soil, damaging plants and contaminating underground aquifers. This pollution can persist for years, making it difficult for affected ecosystems to recover.

4. Impact on Human Health

Oil spills can have serious repercussions on human health. People who live near affected areas may be exposed to toxic fumes, which can cause respiratory problems, skin irritation, and other health problems. In addition, contamination of water sources can lead to gastrointestinal problems and diseases.

5. Economic Consequences

Oil spills also have a significant economic impact. The fishing and tourism industries are usually the most affected. Water pollution can lead to fishing bans, which affects fishermen's livelihoods. Biodiversity loss and the degradation of natural landscapes can discourage tourism, causing additional economic losses.

Significant Examples of Oil Spills

Throughout history, there have been several significant oil spills that have had a lasting impact on the environment and human communities. Below are some of the most notable examples:

1. Exxon Valdez Spill (1989)

One of the most infamous oil spills in history occurred on March 24, 1989, when the Exxon Valdez tanker ran aground in Prince William Strait, Alaska. The accident released approximately 11 million gallons of crude oil into the waters of the strait, causing an ecological catastrophe.

Environmental impact

The Exxon Valdez spill devastated marine and coastal ecosystems. It is estimated that hundreds of thousands of seabirds, thousands of marine mammals and millions of fish died. Sensitive habitats, such as salt marshes and intertidal zones, were covered in oil, affecting plant and animal communities.

Response and Cleaning

The initial response to the spill was slow and often ineffective. Cleanup techniques included the use of chemical dispersants, controlled oil flaring, and manual beach cleaning. However, many of these methods had negative side effects and failed to completely remove the oil.

Legacy

The Exxon Valdez spill led to significant changes in the regulation of oil transportation in the United States, including the implementation of the Oil Pollution Act of 1990, which improved spill prevention and response measures. Despite these efforts, the affected ecosystems took decades to recover and some damage persists to this day.

2. Deepwater Horizon Spill (2010)

The Deepwater Horizon spill is considered one of the worst ecological disasters in history. On April 20, 2010, the Deepwater Horizon oil rig, operated by BP, suffered an explosion in the Gulf of Mexico. The accident resulted in the deaths of 11 workers and the release of approximately 210 million gallons of oil into the ocean over a period of 87 days.

Environmental impact

The Deepwater Horizon spill had a devastating impact on the Gulf of Mexico. The oil affected vast areas of marine ecosystems, including salt marshes, mangroves, and coral reefs. Marine wildlife, including turtles, dolphins and seabirds, suffered mass deaths and long-term effects on health and reproduction.

Response and Cleaning

The response to the spill involved thousands of workers and a variety of techniques, including the application of chemical dispersants, mechanical oil gathering, and controlled burning. Despite these efforts, large amounts of oil remained in the environment, and the effects of the spill are estimated to persist for decades.

Legacy

The Deepwater Horizon spill led to a global reassessment of safety in deepwater drilling. In the United States, new regulations were implemented to improve safety and spill response. However, the disaster also highlighted the limitations of technology and human capacity to handle spills of this magnitude.

3. Prestige Spill (2002)

The Prestige was an oil tanker that sank off the coast of Galicia, Spain, in November 2002, releasing approximately 77,000 tons of oil into the Atlantic Ocean. This disaster had a devastating impact on the northwest coast of Spain and local economies.

Environmental impact

The oil spilled by the Prestige affected more than 1,000 kilometers of coastline, including beaches, marshes and cliffs. Marine wildlife, including birds and fish, suffered mass die-offs, and coastal ecosystems took years to recover.

Response and Cleaning

The response to the spill included collecting oil at sea, manually cleaning beaches and applying dispersants. However, adverse weather conditions and the large extent of the spill made cleanup efforts difficult.

Legacy

The Prestige spill led to changes in shipping regulation in Europe, including a ban on single-hull tankers in European waters. The disaster also highlighted the need to improve spill prevention and response measures internationally.

Conclusion

Oil spills are one of the most visible and devastating manifestations of oil's environmental impact. Not only do these events cause immediate and severe damage to ecosystems and wildlife, but they also have long-lasting effects on human health and local economies. Each of the examples presented demonstrates the magnitude of the consequences of these disasters and the importance of implementing effective measures to prevent and mitigate oil spills.

As the world moves towards a more sustainable energy future, it is crucial to learn from these disasters and work to prevent them from happening again. This involves improving regulation and oversight of the oil industry, developing safer and more efficient technologies, and promoting a transition to cleaner and more sustainable energy sources. Only through a concerted effort can we protect our planet and ensure a healthier and safer future for generations to come.

Air pollution and its consequences

The combustion of oil and its derivatives is not only essential for our daily energy needs, but it is also one of the main sources of air pollution. The effects of this pollution are felt globally, impacting human health, the environment, and the climate. In this section, we will explore how burning oil pollutes the air, the consequences of this pollution, and efforts to mitigate these impacts.

What is Air Pollution?

Air pollution refers to the presence of harmful substances in the atmosphere that can have adverse effects on human health, ecosystems and the climate. These substances include gases, particles, and other pollutants that come from various sources, including the burning of fossil fuels such as petroleum.

Main Pollutants Emitted by Oil Burning

The combustion of oil and its derivatives releases a number of pollutants into the air, each with different detrimental effects:

1. Carbon Dioxide (CO_2): It is the main greenhouse gas emitted by burning oil. Although CO_2 is not toxic in normal concentrations, its accumulation in the atmosphere contributes to global warming and climate change.

2. Nitrogen Oxides (NOx): These gases are products of high-temperature combustion. NOx can irritate the human respiratory system and contribute to the formation of smog and acid rain.

3. Carbon Monoxide (CO): This colorless, odorless gas is toxic to humans and animals. It is formed by the incomplete combustion of fossil fuels and can cause serious health problems, including death in high concentrations.

4. Volatile Organic Compounds (VOCs): These include a variety of chemicals that evaporate easily in the air. VOCs contribute to the formation of tropospheric ozone (ground-level ozone), a major component of smog.

5. Particulate Matter (PM): Burning oil produces fine and ultrafine particles that can penetrate deep into the lungs and enter the bloodstream, causing respiratory and cardiovascular diseases.

6. Sulfur Dioxide (SO_2): This gas is a byproduct of burning sulfur-containing oil. SO_2 can irritate the respiratory system and contribute to the formation of acid rain.

Consequences of Air Pollution

Air pollution from burning oil has a number of negative consequences that affect both humans and the environment. These consequences can be immediate or long-term and vary in severity and scope.

Impact on Human Health

1. Respiratory Diseases: Exposure to air pollutants such as NOx, SO_2 and particulate matter (PM) can cause or aggravate respiratory diseases such as asthma, chronic bronchitis and emphysema. Children, the elderly, and people with pre-existing conditions are especially vulnerable.

2. Cardiovascular Disease: Air pollution is linked to an increase in rates of cardiovascular disease. Fine particles can enter the bloodstream through the lungs, causing inflammation and damaging blood vessels, which can lead to heart attacks and strokes.

3. Cancer: Some air pollutants, such as volatile organic compounds (VOCs) and benzene, are known carcinogens. Long-term exposure to these pollutants increases the risk of developing cancer, especially lung cancer.

4. Child Development Problems: Exposure to air pollutants during pregnancy and childhood can affect the development of children's nervous systems and lungs, resulting in long-term health problems.

5. Premature Mortality: Air pollution is a significant cause of premature mortality worldwide. The World Health Organization (WHO) estimates that air pollution is responsible for millions of premature deaths each year.

Impact on the Environment

1. Acid Rain: Nitrogen oxides (NOx) and sulfur dioxide (SO_2) can react with water, oxygen, and other chemicals in the atmosphere to form acids that then fall to earth in the form of acid rain. Acid rain can acidify soils and bodies of water, damaging terrestrial and aquatic ecosystems.

2. Photochemical Smog: Smog is a mixture of air pollutants, mainly ground-level ozone, that is formed when volatile organic compounds (VOCs) and nitrogen oxides (NOx) react under sunlight. Smog reduces visibility and can cause damage to plants, affecting agricultural production and natural ecosystems.

3. Damage to Vegetation: Air pollution can directly harm plants, affecting their growth and development. Tropospheric ozone, for example, can cause lesions in leaves, reducing the ability of plants to photosynthesize and produce food.

4. Affectation of Fauna: Animals, especially those that live near industrial or urban areas, can also be affected by air pollution.

Pollutants can build up in their bodies, affecting their health and ability to reproduce.

Impact on Climate

1. Climate Change: Carbon dioxide (CO_2) and other greenhouse gases emitted by burning oil contribute to global warming. These gases trap heat in the atmosphere, leading to an increase in global temperatures, melting glaciers, and rising sea levels.

2. Altering Weather Patterns: Air pollution can alter regional and global weather patterns. For example, suspended particles (aerosols) can influence cloud formation and precipitation, affecting hydrological cycles and weather conditions.

3. Ocean Acidification: The increase in CO_2 in the atmosphere also dissolves in the oceans, causing acidification. This affects marine organisms, especially those with shells or skeletons made of calcium carbonate, such as corals and some mollusks.

Efforts to Mitigate Air Pollution

Globally, various efforts are being made to reduce air pollution and its consequences. These efforts include government regulations, technological advances, and changes in industrial and consumer practices.

Regulations and Policies

1. Emission Standards: Many countries have implemented strict emission standards for vehicles and industrial plants. These standards limit the amount of pollutants that can be released into the air, incentivizing the use of cleaner technologies.

2. International Agreements: Treaties such as the Kyoto Protocol and the Paris Agreement seek to reduce global greenhouse gas emissions. These agreements promote international cooperation and set emission reduction targets for participating countries.

3. Renewable Energy Policies: Promoting renewable energy sources, such as solar, wind, and hydropower, helps reduce dependence on oil and other fossil fuels. Many governments offer incentives for the adoption of clean technologies.

Technological Advances

1. Carbon Capture and Storage (CCS) Technologies: These technologies capture CO_2 emitted by industrial and power plants before it reaches the atmosphere, storing it safely in underground geological formations.

2. Electric Vehicles: The transition to electric vehicles significantly reduces NOx, CO and particulate matter emissions, as these vehicles do not burn fossil fuels. Improvements in charging infrastructure and battery efficiency are accelerating this transition.

3. Energy Efficiency Improvements: The implementation of more efficient technologies in power generation, building construction, and industry can significantly reduce energy consumption and pollutant emissions.

Changes in Industrial and Consumer Practices

1. Public Transport and Sustainable Mobility: Encouraging the use of public transport, bicycles and walking can reduce the number of vehicles on the road and, therefore, pollutant emissions.

2. Responsible Consumption: Adopting responsible consumption habits, such as reducing the use of plastics and disposable products, can decrease the demand for petroleum products and, consequently, pollutant emissions.

3. Reforestation and Conservation: Planting trees and conserving forests and other natural ecosystems can help absorb CO_2 from the atmosphere, mitigating climate change.

Conclusion

Air pollution caused by oil burning is a global problem with serious and far-reaching consequences. It affects human health, harms the environment, and contributes to climate change. However, through regulations, technological advances, and changes in industrial and consumer practices, it is possible to mitigate these impacts.

As the world moves towards a more sustainable future, it is crucial to continue to develop and adopt solutions that reduce our dependence on oil and minimise air pollution. Not only will this improve the quality of the air we breathe, but it will also protect our health, our ecosystems, and our climate for future generations. The transition to a cleaner, healthier world is a monumental challenge, but with global cooperation and commitment, it is an achievable goal.

The challenge of waste and water pollution

Petroleum, in addition to being an important source of energy, also generates a considerable amount of waste and contributes significantly to water pollution. From extraction to refining to consumption, every stage of the petroleum life cycle has the potential to pollute our water resources and produce hazardous waste. This section examines these challenges and the environmental consequences they entail in detail.

Waste from the Oil Industry

The oil industry generates various types of waste throughout its supply chain. This waste can be solid, liquid or gaseous and often contains toxic substances that can harm the environment and human health.

1. Solid Waste

Solid waste produced by the petroleum industry includes:

- Drilling Debris: During the drilling of oil wells, large amounts of debris are generated. These can contain toxic materials, such as heavy metals and chemicals used in the drilling process.

- Refining Sludge: Oil refining produces sludge and sludge that contain hydrocarbons and other contaminants. This waste is often difficult to handle and requires special treatments.

- Maintenance Waste: Equipment and platform maintenance operations generate solid waste, such as used filters, oil-impregnated rags, and obsolete or damaged equipment parts.

2. Liquid Waste

Liquid waste is perhaps the most problematic due to its ability to contaminate water sources:

- Drilling Wastewater: The drilling process uses large amounts of water mixed with chemicals to lubricate the drill bit and bring the debris to the surface. This wastewater can contain toxic substances that, if not properly managed, can leach into the soil and contaminate groundwater.

- Production Water: During oil extraction, groundwater containing hydrocarbons and dissolved salts is also extracted. This production water must be treated before being discarded or reused.

- Refining Effluents: Oil refineries generate liquid effluents that contain hydrocarbons, heavy metals, and other pollutants. These effluents need to be treated to reduce their environmental impact before being discharged into bodies of water.

3. Gaseous emissions

Although gaseous emissions do not directly pollute water, they contribute to climate change and air pollution, which can have indirect effects on water resources:

- Flue Gases: The burning of oil and its derivatives in industrial processes and internal combustion engines releases gases such as carbon dioxide (CO_2), nitrogen oxides (NOx) and sulfur dioxide (SO_2).

- Methane leaks: During the extraction and transportation of oil, leaks of methane, a potent greenhouse gas, can occur.

Oil Water Pollution

Oil and its derivatives can pollute water in a variety of ways, from accidental spills to chronic leaks and intentional discharges. Oil water pollution has devastating effects on aquatic ecosystems and human health.

1. Oil Spills

Oil spills are perhaps the most visible form of water pollution. These events can occur at sea or on land, and the impacts are always severe:

- Marine Spills: Spills at sea, such as the Exxon Valdez in Alaska or the Deepwater Horizon in the Gulf of Mexico, release large amounts of oil into the water, affecting vast areas of oceans and coastlines. The oil spreads rapidly, forming a layer that blocks sunlight and affects the photosynthesis of aquatic plants. In addition, the oil attaches to marine wildlife, causing the death of birds, mammals, and fish.

- Land-based spills: Land-based spills can contaminate rivers, lakes, and underground aquifers. The spilled oil infiltrates the soil, affecting plant roots and groundwater. This can have long-lasting effects on agriculture and drinking water supplies.

2. Leaks and Downloads

Even without catastrophic spills, the oil industry can pollute water through seepage and discharge:

- Pipeline Leaks: Pipelines that transport oil can leak due to corrosion, seismic movements, or damage from human activities. These leaks release oil into soil and groundwater, contaminating drinking water sources and affecting aquatic life.

- Intentional discharges: In some regions, oil companies intentionally discharge treated or partially treated wastewater into bodies of water. These discharges may contain contaminants that affect water quality and the health of aquatic ecosystems.

3. Refinery Effluents

Oil refineries produce large quantities of effluents containing hydrocarbons, heavy metals, and other pollutants. These effluents, if not treated properly, can contaminate nearby bodies of water:

- Chemical Pollution: Chemicals used in petroleum refining, such as solvents and catalysts, can be toxic to aquatic life. Prolonged exposure to these pollutants can cause the death of fish and other aquatic life.

- Impact on Biodiversity: Effluents from refineries can alter the chemical composition of water, affecting the biodiversity of aquatic ecosystems. The most sensitive species may disappear, while others, less susceptible, may proliferate, altering the ecological balance.

Consequences of Water Pollution

Water pollution by oil and its derivatives has serious and far-reaching consequences that affect ecosystems, human health and local economies.

1. Impact on Aquatic Ecosystems

Water pollution can devastate aquatic ecosystems, affecting all forms of life that depend on these environments:

- Toxic Effects on Wildlife: Hydrocarbons and other pollutants present in oil are toxic to many aquatic species. Fish, crustaceans, mollusks, and other marine organisms can suffer organ damage, reproductive problems, and mass mortality.

- Biodiversity Loss: Chronic pollution and massive spills can lead to the loss of biodiversity in aquatic ecosystems. The most vulnerable species may disappear, affecting the food chain and ecological balance.

- Damage to Habitats: Mangroves, coral reefs, and salt marshes are particularly sensitive to oil pollution. These habitats are crucial for many species, and their destruction can have cascading effects on marine ecosystems.

2. Impact on Human Health

Oil water pollution also has serious implications for human health:

- Contaminated Drinking Water: Oil infiltration into aquifers and other drinking water sources can make water unsafe for human consumption. The toxic compounds present in oil can cause a variety of health problems, including cancer, liver and kidney disease, and neurological disorders.

- Exposure to Contaminants: Communities that rely on fishing and other water-related activities may be exposed to contaminants through the consumption of contaminated fish and shellfish.

Polycyclic aromatic hydrocarbons (PAHs), present in petroleum, are particularly dangerous due to their carcinogenic properties.

3. Economic Impact

Oil water pollution also has significant economic consequences:

- Fishing Industry: Pollution of water bodies can affect the fishing industry, leading to loss of income for fishermen and coastal communities. Declining fish and shellfish populations due to contamination affect product availability and quality.

- Tourism: Coastal areas affected by oil spills and pollution may see a decline in tourism. Polluted beaches, the death of marine life and the negative perception of the area can discourage tourists, affecting local economies.

- Cleanup and Remediation Costs: The costs associated with cleaning up and remediating oil spills and water contamination can be extremely high. These costs are often borne by governments and, ultimately, by taxpayers.

Efforts to Mitigate Oil Water Pollution

Given the significant impact of oil water pollution, various efforts have been implemented to mitigate these effects. These efforts include regulations, advanced technologies, and improved practices in the petroleum industry.

1. Regulations and Policies

Governments and international organizations have established a number of regulations and policies to reduce water pollution from oil:

- Emissions and Discharge Regulations: Many countries have implemented strict regulations on emissions and discharges from the oil industry. These regulations limit the amount of pollutants that can be released into the environment and establish procedures for the safe handling of waste.

- International Treaties: Treaties such as the International Convention on Oil Pollution Preparedness, Response and Cooperation (OPRC) and the MARPOL Convention seek to prevent marine pollution from ships and oil platforms. These treaties promote international cooperation and set standards for oil spill prevention and response.

2. Advanced Technologies

Advanced technologies play a crucial role in preventing and mitigating oil water pollution:

- Detection and Monitoring Technologies: Modern technologies allow for the early detection of leaks and spills. Advanced sensors, drones, and satellites can monitor oil facilities and transportation routes, providing early warnings and accurate data for rapid response.

- Containment and Recovery Systems: Containment systems, such as floating booms and skimmers, can limit the spread of spilled oil and facilitate its recovery. These technologies are essential to minimize the impact of spills on water bodies.

- Wastewater Treatment: Wastewater treatment plants can effectively remove contaminants from production water and refinery effluent. Membrane technology, advanced filtration, and biological processes are some of the techniques used to treat these waters.

3. Improved Practices in the Oil Industry

The oil industry is adopting improved practices to minimize its environmental impact:

- Waste Management: Proper management of solid and liquid waste is crucial to reducing pollution. This includes waste separation and treatment, recycling of materials, and safe disposal of hazardous waste.

- Infrastructure Maintenance and Upgrade: Regular maintenance and upgrading of infrastructure, such as pipelines and platforms, can prevent leaks and spills. Implementing modern technologies and improving safety procedures are essential to minimize risks.

- Training and Awareness: Ongoing worker training and awareness of environmental practices are critical to ensuring that all employees follow safety and environmental management protocols. Oil companies are investing in training and awareness programs to promote a culture of safety and environmental responsibility.

Conclusion

The challenge of waste and water pollution associated with the petroleum industry is significant and multifaceted. From solid and liquid waste generation to catastrophic spills and chronic leaks, every stage of the oil life cycle presents environmental risks that must be carefully managed.

Oil water pollution has devastating effects on aquatic ecosystems, human health, and local economies. However, through strict regulations, the use of advanced technologies, and the adoption of improved industry practices, it is possible to mitigate these impacts and protect our valuable water resources.

As the world moves towards a more sustainable future, it is crucial to continue to develop and adopt innovative solutions that reduce dependence on oil and minimize water pollution. Protecting our water resources is not only vital to the health of the planet, but also to the survival and well-being of all forms of life that depend on them. With global commitment and concerted action, we can meet this challenge and ensure a cleaner, healthier future for future generations.

Chapter 4: Time for Change: The Electric Vehicle on Stage

The history of the electric car: From the first prototypes to the present day

The electric vehicle, while it may seem like a recent innovation, has a fascinating history dating back more than a century. From the first prototypes to today's advanced models, electric cars have come a long way. This evolution has been driven by technological advances, changes in energy policies, and a growing awareness of the need for more sustainable transportation. Below, we explore the rich history of the electric car, highlighting its key moments and the milestones that have marked its development.

The Early Days: Nineteenth Century

Early Beginnings and Prototypes

The origin of the idea of an electric-powered vehicle dates back to the nineteenth century, a time of great technological development. Between 1828 and 1835, the first experiments were carried out: in 1828, Ányos Jedlik, a Hungarian innovator, invented a small electric motor that he used to move a model vehicle. In 1834, American Thomas Davenport manufactured an electric motor that powered a small automobile, marking one of the first efforts to use electricity for transportation. In 1835, Scottish inventor Robert Anderson created the first rudimentary electric car with a non-rechargeable battery, a significant advance in the history of the electric car.

Development of Batteries and Vehicles (1860-1890)

The development of rechargeable batteries was crucial to the advancement of electric cars.

In 1859, French physicist Gaston Planté invented the lead-acid battery. This rechargeable battery provided a more efficient and reliable source of power for electric vehicles. Subsequently, in 1881, Camille Alphonse Faure improved Planté's lead-acid battery, increasing its capacity and durability, which made the batteries more practical for use in vehicles.

As for the first commercial vehicles, in 1881, the French inventor Gustave Trouvé presented an electric tricycle at an exhibition in Paris, powered by a lead-acid battery. This was one of the first examples of an electric car in operation. In 1884, British engineer Thomas Parker built one of the first practical electric vehicles, using high-capacity batteries, and is considered a pioneer in the electrification of transportation.

The Golden Age of Electric Vehicles: 1890-1920

In the late 19th and early 20th centuries, electric vehicles enjoyed a golden age. During this period, they competed favorably with gasoline and steam cars.

Initial Advantages of Electric Vehicles

Electric cars offered a number of advantages that made them attractive to users. First, they were more comfortable and easier to use than gasoline cars. While gasoline-powered vehicles required hand starting and continuous engine adjustments, electric cars were started with the turn of a key, making them especially popular with women and those seeking comfort.

In addition, electric cars were quiet and clean, unlike noisy and polluting gas-powered cars. They produced no exhaust emissions, making them ideal for use in urban areas.

Thanks to these features, electric cars became common in cities such as New York, London, and Paris. They were used as taxis, luxury cars and delivery vehicles, consolidating their popularity in urban environments.

Innovations and Classic Models

In 1898, Austrian engineer Ferdinand Porsche developed the P1, one of the first hybrid electric cars. This vehicle could run on both electricity and petrol, demonstrating the versatility and potential of electric powertrains.

The Baker Motor Vehicle Company of Cleveland, Ohio, produced the Baker Electric between 1899 and 1916. This electric car stood out for its sleek design and popularity among the wealthy elite, as well as offering a reasonable range for its time.

For its part, the Anderson Electric Car Company manufactured the Detroit Electric between 1907 and 1939, making it one of the most successful electric cars of its time. With a range of up to 80 miles on a single charge, this vehicle was practical for urban driving, cementing the viability of electric cars in city environments.

These pioneers and their innovations laid the foundation for the development of electric vehicle technology, showcasing its potential and appeal in various market segments.

Temporary Decline of Electric Vehicles

Despite their initial success, several factors contributed to the decline of electric vehicles starting in the 1920s. One of the main reasons was significant improvements in internal combustion engines. Henry Ford led the mass production of gasoline-powered cars with his Model T, making these vehicles more affordable and accessible to the general public.

The expansion of gasoline supply infrastructure also played a crucial role. The widespread availability of petrol stations made it easier to use petrol cars. In addition, the invention of electric starting for internal combustion engines eliminated one of the main advantages of electric cars, which was ease of starting.

On the other hand, the limitations of lead-acid batteries in terms of weight, capacity, and recharging time made electric vehicles less competitive. These batteries were heavy and offered limited range, making electric cars unable to compete with the convenience and efficiency of gasoline-powered vehicles.

In short, the combination of technological advances in gasoline engines, the expansion of fuel infrastructure, and the limitations of lead-acid batteries led to the decline of electric vehicles in the 1920s. Renaissance and Evolution: 1960-2000

After several decades of neglect, electric vehicles began to re-emerge in the second half of the 20th century, driven by concerns about the environment and the energy crisis.

Oil Crisis of the 1970s

The 1973 oil crisis highlighted the vulnerability of oil dependence and stimulated interest in alternative sources of energy. This critical event led governments and companies to invest in the research and development of electric vehicles. The crisis increased awareness of the need to diversify energy sources and reduce dependence on imported oil.

During the 1970s, several manufacturers launched experimental models of electric cars. These vehicles, although limited in range and speed, represented a significant step towards the adoption of cleaner transport technologies. Among these experimental models was the Sebring Vanguard CitiCar, a small electric car that gained popularity due to growing concerns about oil shortages. This car, along with other similar projects, demonstrated the potential of electric vehicles and laid the foundation for future developments in this field.

Battery research also saw breakthroughs during this period. Alternatives to lead-acid batteries, such as nickel-cadmium batteries and nickel-iron batteries, were explored, although these also had their own limitations. Despite the technological challenges, the 1970s marked a turning point in the history of electric vehicles, paving the way for the innovations that would come in the decades that followed.

Advances in Battery Technology

The development of new battery technologies was crucial to the evolution of electric vehicles.

Over the decades, nickel-cadmium (NiCd) and nickel-metal-hydride (NiMH) batteries offered significant improvements in energy density and duration compared to lead-acid batteries. These batteries made it possible to increase the range and improve the performance of electric vehicles, making them more viable for everyday use. NiMH batteries, in particular, stood out for their ability to store more energy and withstand more frequent charge and discharge cycles without losing efficiency.

In the 1990s, several manufacturers launched electric cars equipped with NiMH batteries. A prominent example is the GM EV1, launched by General Motors in 1996. This was one of the first modern electric vehicles to be mass-marketed and symbolized a breakthrough in the automotive industry. The GM EV1 offered a range of up to 160 kilometers per charge, which was remarkable for the time and showed the potential for electric vehicles to replace internal combustion vehicles in certain market segments. In addition to the GM EV1, other manufacturers also explored the use of NiMH batteries in their electric models, which further boosted research and development in this area.

These technological advances in batteries laid the foundation for the next generation of electric vehicles, paving the way for the advent of lithium-ion batteries and other developments that have led to the modern era of electric mobility.

Regulations and Policies

Government regulations also played a crucial role in the revival of electric vehicles.

In 1990, California introduced the Zero-Emission Vehicle (ZEV) Mandate, which required automakers to produce a certain percentage of zero-emission vehicles. This regulation was a significant catalyst for the development of electric cars and other clean transportation technologies. The ZEV mandate prompted major automakers to invest in EV research and development, resulting in models such as the GM EV1 and Toyota RAV4 EV, both pioneers of their time.

In addition to the ZEV mandate, several governments implemented incentives and subsidies to encourage the adoption of electric vehicles. These incentives included tax breaks for EV buyers, purchase grants, and programs for the installation of charging stations. In the United States, the federal tax credit offered up to $7,500 for the purchase of an electric vehicle, making these cars more accessible to the public. In Europe, countries such as Norway and Germany offered tax breaks and generous subsidies, resulting in rapid adoption of electric vehicles.

These government efforts not only helped reduce greenhouse gas emissions, but also spurred innovation in the automotive sector, accelerating the development of advanced battery technologies and the charging infrastructure needed to support the transition to more sustainable transportation.

The 21st Century: The Age of the Electric Vehicle

The 21st century has seen a significant resurgence and widespread acceptance of electric vehicles, driven by technological advances, favorable policies, and growing environmental awareness.

Tesla and the Electric Vehicle Revolution

Tesla Motors, founded in 2003, has been a transformative force in the electric car industry. Its impact began with the launch of the Tesla Roadster in 2008, the company's first production car. Equipped with a lithium-ion battery, the Roadster proved that electric cars could be fast and have a considerable range, with a top speed of 201 km/h and a range of approximately 393 km per charge. This initial success helped change the public perception of electric vehicles, proving that they could compete with gasoline cars in terms of performance and range.

The next major milestone for Tesla was the launch of the Model S in 2012. This luxury sedan offered a combination of range, performance, and advanced features that appealed to a broad spectrum of consumers. With a range of up to 647 km, the Model S set new standards for electric vehicles. Tesla subsequently launched the Model X, an SUV with hawk-wing doors and advanced safety features, and the Model 3, a more affordable sedan designed for the mass market. These models cemented Tesla as a leader in the electric vehicle market, offering options that combined luxury, innovation, and sustainability.

To support the adoption of its vehicles, Tesla invested significantly in creating a network of superchargers. This fast-charging infrastructure allows Tesla owners to conveniently and efficiently charge their vehicles around the world. By the end of 2020, Tesla had installed more than 20,000 superchargers, addressing one of the main obstacles to EV adoption: the availability of charging stations. Not only does this network facilitate long journeys, but it also reduces range anxiety, a key factor in the widespread adoption of electric vehicles.

In addition to its vehicles and charging infrastructure, Tesla has pioneered other areas of technology, such as developing advanced software for autonomous driving and integrating renewable energy into its products. Elon Musk's vision for a sustainable future has led the company to continuously innovate and challenge automotive industry norms, positioning Tesla not only as a carmaker, but as a leader in the transition to a cleaner and more sustainable world.

Advances in Battery Technology

The development of lithium-ion batteries has been instrumental in improving the performance of electric vehicles, representing a significant technological advance.

Lithium-ion batteries offer higher energy density compared to previous technologies, such as lead-acid and nickel-cadmium batteries. This higher energy density allows electric vehicles to have a longer range and lower weight, which improves vehicle performance and efficiency. For example, while lead-acid batteries had an energy density of around 30-50 Wh/kg, modern lithium-ion batteries can achieve densities of 150-250 Wh/kg, allowing electric cars to travel longer distances on a single charge.

In addition to the improvement in energy density, cost reduction has been a crucial factor for the mass adoption of electric vehicles. Over the years, the cost of lithium-ion batteries has dropped dramatically. In 2010, the average cost of batteries was about $1,100 per kWh. By 2020, this cost had dropped to around $137 per kWh, according to BloombergNEF. This decrease in costs has made electric vehicles more affordable for the average consumer, allowing more people to consider buying an electric car as a viable option.

These advancements have been made possible by continuous research and development in the field of batteries. Innovations such as high-capacity cathodes and advanced electrolytes have improved the efficiency and safety of lithium-ion batteries, while economies of scale and optimization of manufacturing processes have contributed to cost reductions. Together, these developments have driven the adoption of electric vehicles, marking a significant milestone in the transition to cleaner and more sustainable transport.

Competition and Diversification in the Market

Tesla's success spurred other automakers to develop and launch their own electric vehicles, diversifying the market and offering more choice to consumers.

In 2010, Nissan launched the Leaf, one of the first mass-market electric vehicles accessible. Equipped with a 24 kWh battery, the Leaf offered a range of around 117 kilometers, which made it popular with urban drivers. Its success helped demonstrate the commercial viability of electric cars and increased their public acceptance, surpassing sales of 500,000 units globally by 2020. Nissan continued to improve the Leaf, increasing its range and adding advanced features such as ProPilot autonomous driving.

Other manufacturers also followed this path. BMW launched the i3 in 2013, an electric vehicle with an innovative design and a carbon fiber structure, offering a range of up to 200 kilometers. The i3 was notable not only for its efficiency, but also for its focus on sustainability, using recycled materials and a low-emission manufacturing process. Chevrolet introduced the Bolt in 2016, an electric car with a 60 kWh battery and a range of approximately 383 kilometers, positioning itself as a direct competitor to Tesla in terms of performance and range. The Bolt was well received for its combination of affordable price, range, and technological features.

Renault also contributed to the market with the Zoe, which became one of the best-selling electric vehicles in Europe. The Zoe offered competitive range and an affordable price, which made it attractive to a wide range of consumers. Renault expanded its offering with upgraded versions of the Zoe, increasing battery capacity and range, and adding advanced connectivity and safety features.

The entry of these and other manufacturers into the electric vehicle market has led to rapid innovation and expansion of the electric model offering. Brands such as Audi, Jaguar and Porsche have launched their own luxury electric vehicles, while traditional manufacturers such as Ford and Volkswagen are investing heavily in the development of electric platforms and the electrification of their product lines.

In addition, the competition has fostered continuous improvements in battery technology, the efficiency of electric motors and charging infrastructure. Consumers now have a wide range of options ranging from small urban vehicles to SUVs and luxury sedans, all with different levels of range, performance and technological features. This diversification has allowed electric vehicles to adapt to the needs and preferences of an ever-wider audience, accelerating the transition to a cleaner and more sustainable transport future.

The impact of Tesla and other pioneers in the electric vehicle industry has been profound, driving not only the adoption of these vehicles, but also the development of policies and regulations that support the electrification of transportation. With a rapidly evolving market and growing support from governments and consumers, the future of electric vehicles looks brighter than ever.

Global Policies and Regulations

Government policies and regulations have continued to play a crucial role in driving EV adoption.

Many countries have implemented stricter emissions standards, incentivizing manufacturers to develop electric vehicles and other clean technologies to meet these standards. For example, the European Union introduced rules that oblige car manufacturers to reduce the average CO_2 emissions of their fleets, which has boosted the electrification of transport. In 2020, regulations required new vehicles to emit an average of no more than 95 grams of CO_2 per kilometre, a goal that can only be achieved through greater adoption of electric vehicles.

In addition to emissions standards, governments offer financial incentives to encourage the purchase of electric vehicles. These incentives include tax breaks, direct subsidies, and tax breaks. In the United States, the federal tax credit offers up to $7,500 for the purchase of an electric vehicle, while countries like Norway have implemented even more generous policies, such as the VAT exemption, which has led to electric vehicles accounting for more than 50% of new car sales in 2020.

Some countries and cities have announced plans to ban the sale of new internal combustion vehicles in the near future, accelerating the transition to electric transport. For example, the UK has brought forward its ban to 2030, while Norway has set 2025 as the deadline for the sale of new petrol and diesel cars. Not only do these policies drive the adoption of electric vehicles, but they also send a clear signal to automakers about the future direction of the market, incentivizing investment in clean and sustainable technologies.

These policies and regulations are supported by the development of charging infrastructure and investment in research and development of battery and electric vehicle technologies. The combination of strict regulations, financial incentives, and future bans is creating a favorable environment for the mass adoption of electric vehicles, propelling the industry towards a cleaner and more sustainable future.

Charging Infrastructure and Assistive Technologies

The expansion of charging infrastructure and the development of supporting technologies are crucial for the widespread adoption of electric vehicles.

Installing fast-charging stations in urban areas, highways, and popular destinations makes it easy to use electric vehicles for long trips and daily commutes. For example, Tesla has developed its supercharger network that allows its users to recharge up to 80% of the battery in approximately 30 minutes. In addition, companies such as Electrify America and Ionity are installing fast-charging stations around the world to support the growing demand for electric vehicles.

The availability of home and workplace chargers is also essential. Home chargers allow EV owners to conveniently charge their cars overnight, using lower electricity rates. In the workplace, chargers allow employees to recharge their vehicles while working, increasing the convenience and practicality of everyday electric car use. According to a report by the International Energy Agency (IEA), most EV charging is done at home or at work, underscoring the importance of these charging points.

The continuous development of new battery technologies promises to further improve the performance of electric vehicles. Solid-state batteries, for example, offer higher energy density, faster charging times, and greater safety compared to current lithium-ion batteries. Toyota and other manufacturers are investing in this technology, with plans to introduce vehicles equipped with solid-state batteries in the next decade. These innovations could allow electric vehicles to travel longer distances on a single charge and significantly reduce the time needed to recharge.

In summary, the combination of fast-charging infrastructure, the availability of home and on-the-job chargers, and continuous innovations in battery technology are key factors for the mass adoption of electric vehicles. These advances not only improve the convenience and viability of electric vehicles, but also contribute to a more sustainable and emission-free future.

Future Prospects

The future of electric vehicles looks very promising, with continued technological advancements, supportive policies, and growing public acceptance.

Automakers are significantly increasing EV production and diversifying their offerings to include a variety of models, from compacts to SUVs to trucks. Brands like Volkswagen, Ford and General Motors are investing billions in electrifying their fleets, with plans to launch dozens of new electric models in the next decade. This expansion not only increases the availability of electric vehicles, but also provides more options for consumers, adapting to different needs and preferences.

The integration of electric vehicles with renewable energy sources, such as solar and wind, is another important development. Companies like Tesla and Nissan are developing solutions for EV owners to charge their cars with home solar power. In addition, connecting these vehicles to smart grids enables efficient energy management, using electricity generated by renewable sources for charging and storing the surplus in car batteries, thus contributing to grid stability and reducing the carbon footprint.

Autonomous driving technologies and improvements in the range of electric vehicles also promise to transform the way we get around. Advances in artificial intelligence and sensors have allowed companies such as Waymo, Tesla and GM Cruise to develop autonomous driving systems that are being tested in various cities around the world. These systems can improve road safety by reducing human error and increasing traffic efficiency. At the same time, improvements in batteries, such as the development of solid-state batteries, are increasing the range of electric vehicles, allowing longer journeys with fewer recharges.

In short, the future of electric vehicles is bright, driven by a combination of diversified production, integration with renewable energy and advanced autonomous driving technologies. Not only are these developments making electric vehicles more accessible and practical, but they are also helping to create a more sustainable and efficient transportation system.

Conclusion

The history of the electric car is a testament to innovation and resilience in the pursuit of cleaner and more sustainable transport. From the first prototypes in the 19th century, such as Gustave Trouvé's electric tricycle and Ferdinand Porsche's P1, to today's advanced models, electric vehicles have evolved significantly. This journey has been driven by technological advances, such as the invention of lithium-ion batteries in the 1990s, which offered higher energy density and lower weight, and changes in energy policies, including government incentives and stricter emissions regulations, that have encouraged the adoption of clean technologies.

The 21st century has seen a significant resurgence of electric vehicles, led by companies such as Tesla, which with the launch of the Roadster in 2008 demonstrated that electric cars could combine performance and range. The expansion of charging infrastructure, with supercharger networks and fast-charging stations, has been crucial to widespread adoption. Integration with renewable energies, such as solar and wind, is also making transport even more sustainable.

As we move into the future, electric vehicles are destined to play a central role in transforming the automotive industry and fighting climate change. Innovations in battery technologies, such as solid-state batteries, and advances in autonomous driving promise to improve transportation efficiency and safety. With continuous support and constant innovation, the electric car is not only the vehicle of the future, but also a tangible solution to the environmental challenges of our time. This evolution reflects the global commitment to sustainability and environmental protection.

Environmental advantages of the electric car

Electric vehicles (EVs) represent a crucial innovation in the pursuit of more sustainable and environmentally friendly transport. By reducing or eliminating dependence on fossil fuels, electric cars offer numerous environmental benefits that are essential for combating climate change, improving air quality, and preserving natural resources. In this section, we explore in depth the environmental benefits of electric cars and how they contribute to a greener future.

1. Reduction of Greenhouse Gas Emissions

One of the most significant advantages of electric vehicles is their ability to reduce greenhouse gas (GHG) emissions, which are primarily responsible for climate change.

CO_2 emissions

Internal combustion vehicles (ICVs) burn gasoline or diesel, releasing carbon dioxide (CO_2) as a byproduct. In contrast, electric vehicles run on electricity and do not emit CO_2 directly during operation. Although electricity generation can produce CO_2 emissions, these are generally lower compared to the emissions of an internal combustion engine, especially when the electricity comes from renewable sources.

It is important to consider emissions throughout the vehicle's life cycle, from production to use and recycling. Studies have shown that, even taking into account the emissions associated with battery production and electricity generation, electric vehicles emit significantly less CO_2 over their lifetime compared to internal combustion vehicles.

Other Greenhouse Gas Emissions

The production and transportation of fossil fuels not only releases carbon dioxide, but also other greenhouse gases such as methane (CH_4) and nitrogen oxides (NOx). Methane is a gas with a much higher global warming potential than CO_2, while nitrogen oxides contribute to the formation of smog and acid rain. By reducing oil demand, electric vehicles not only decrease CO_2 emissions, but also these additional CH_4 and NOx emissions, contributing to better air quality and climate change mitigation.

2. Improved Air Quality

The transition to electric vehicles has a significant positive impact on air quality, especially in densely populated urban areas.

Reduction of Air Pollutants

Internal combustion engines emit nitrogen oxides (NOx), which contribute to the formation of smog and acid rain. Electric vehicles, by not burning fossil fuels, do not emit NOx, which helps to improve air quality.

The combustion of gasoline and diesel also produces fine particulate matter (PM2.5 and PM10) that can penetrate deep into the lungs and cause respiratory and cardiovascular problems. Electric vehicles do not generate these particles directly, reducing air pollution and improving public health.

Carbon monoxide (CO) is a toxic gas that is produced in the incomplete combustion of fossil fuels. By not burning fuel, electric vehicles do not emit CO, contributing to cleaner and healthier air.

Public Health Impact

Exposure to air pollutants such as NOx, PM and CO is linked to a number of health problems, including asthma, bronchitis, heart disease and stroke. Scientific studies have shown that the reduction of these pollutants thanks to electric vehicles can lead to a significant decrease in the incidence of these diseases. A report by the World Health Organization (WHO) highlights that improving air quality can significantly reduce respiratory and cardiovascular problems in the population.

Air pollution is a major cause of premature mortality worldwide. According to the WHO, millions of premature deaths annually are linked to exposure to polluted air. Improving air quality through the adoption of electric vehicles can save lives and improve the quality of life for millions of people. The transition to electric mobility not only has environmental benefits, but also significant positive impacts on global public health, contributing to a decrease in premature mortality and chronic diseases related to air pollution.

3. Superior Energy Efficiency

Electric vehicles are generally more efficient in terms of energy use compared to internal combustion vehicles.

Engine Efficiency

Electric motors convert a higher proportion of the stored energy into motion, with an efficiency of 85-90%, compared to internal combustion engines, which have an efficiency of around 20-30%. This means that electric vehicles use energy more efficiently, reducing the waste of energy resources. Studies by the U.S. Department of Energy indicate that the superior efficiency of electric motors not only optimizes energy consumption, but also contributes to a significant reduction in greenhouse gas emissions when combined with renewable energy sources.

Energy Regeneration

Many electric vehicles are equipped with regenerative braking systems, a technology that recovers kinetic energy that would otherwise be lost during braking and converts it into electricity to recharge the battery. According to the U.S. Environmental Protection Agency (EPA), this system can significantly improve the vehicle's energy efficiency, converting up to 70% of kinetic energy into reusable electrical energy. Not only does this increase the overall efficiency of the vehicle, but it also reduces the need to charge the battery as often, thus extending its lifespan and improving the overall performance of the electric vehicle.

4. Reducing Dependence on Fossil Fuels

The transition to electric vehicles helps reduce the world's dependence on fossil fuels, promoting a more sustainable energy future.

Energy Source Diversification

Electric vehicles can be charged with electricity generated from renewable sources such as solar, wind, and hydropower. According to the International Energy Agency (IEA), as the share of renewable energy in the electricity grid increases, the carbon footprint of electric vehicles decreases further, contributing significantly to the reduction of greenhouse gas emissions.

In addition to renewables, other low-carbon energy sources, such as nuclear power, can also contribute to the electrification of transportation and the reduction of global greenhouse gas (GHG) emissions. Nuclear power, for example, provides a stable, low-carbon source of electricity that can complement intermittent renewable sources, such as solar and wind, ensuring a steady supply of clean energy to charge electric vehicles.

Energy Security

For many countries, dependence on imported oil poses a significant risk to energy security. The adoption of electric vehicles can reduce this dependency, increasing economic and political stability. According to a report by the International Energy Agency (IEA), declining oil demand can protect countries against market fluctuations and geopolitical tensions. This transition not only promotes greater energy independence, but also contributes to economic resilience and a more sustainable and secure energy policy.

5. Noise Pollution Reduction

Electric vehicles contribute to the reduction of noise pollution, especially in densely populated urban areas.

Quiet Operation

Electric motors are significantly quieter than internal combustion engines, reducing noise levels in cities and creating a quieter and more pleasant environment for residents. According to a study by the European Environment Agency (EEA), traffic noise is one of the main sources of noise pollution in urban areas, contributing to health problems such as stress, hypertension and sleep loss. The adoption of electric vehicles can mitigate these negative effects by significantly decreasing the noise associated with urban transportation.

In addition, a report from the UK Department for Transport indicates that electric vehicles can reduce noise levels by up to 4 decibels compared to internal combustion vehicles. This reduction is especially noticeable at low speeds, which is when internal combustion engines are noisiest. In urban environments, where most trips are made at lower speeds, the introduction of electric vehicles can have a considerable impact on decreasing noise pollution.

Reducing traffic noise not only improves the quality of life for urban residents, but also benefits local wildlife, which can be affected by high noise levels. Less noise allows for better communication between animal species and reduces stress on wildlife, contributing to a healthier urban ecosystem.

Impact on Health and Wellbeing

Noise pollution is closely linked to stress, sleep disruption, and other mental health problems. Studies by the World Health Organization (WHO) have shown that prolonged exposure to traffic noise can increase stress levels, cause sleep disturbances, and contribute to the onset of anxiety and depression.

Reducing traffic noise thanks to electric vehicles can significantly improve the quality of life and psychological well-being of people living in urban areas. A report by the European Environment Agency (EEA) indicates that less ambient noise is associated with lower stress rates and better sleep quality, which in turn improves overall well-being. This reduction in noise contributes to a healthier and more peaceful environment, benefiting both the physical and mental health of urban residents.

6. Conservation of Natural Resources

The production and use of electric vehicles can also contribute to the conservation of natural resources.

Reduced Fossil Fuel Extraction

By decreasing the demand for oil, the transition to electric vehicles reduces the need for fossil fuel exploration and extraction. This change helps preserve natural ecosystems and reduces the risk of spills and other environmental disasters. According to a report by the International Energy Agency (IEA), the decrease in oil exploration and extraction protects sensitive areas and reduces habitat destruction. In addition, less drilling and extraction activity decreases the chance of oil spills, which have devastating effects on marine life and coastal ecosystems.

Impact on Mining and Land Use

Although the production of batteries for electric vehicles requires the extraction of minerals such as lithium, cobalt, and nickel, recycling technologies and innovation in battery design are advancing to mitigate these impacts. According to research from the University of California, efficient recycling processes allow up to 95% of used battery materials to be recovered, reducing the need for new extraction. In addition, new battery technologies are being developed, such as solid-state batteries and those that use more abundant materials such as sodium, which are less harmful to the environment. These innovations promise to make electric vehicles even more sustainable in the future.

7. Promotion of Innovation and Sustainable Development

The adoption of electric vehicles is driving innovation in clean technologies and promoting sustainable development.

Development of Clean Technologies

The growing demand for electric vehicles is spurring investment in research and development of new battery technologies, energy management systems, and sustainable materials. This investment translates into innovations that not only benefit the transport sector, but also other industries. For example, advances in solid-state batteries and thermal management systems can improve energy efficiency in consumer electronics and in the renewable energy storage industry. Research from institutions such as the Massachusetts Institute of Technology (MIT) is developing more sustainable battery materials, such as graphene, that offer greater efficiency and lower environmental impact.

Circular Economy

The recycling of EV batteries and components is fostering a circular economy, where materials are recovered and reused rather than discarded. This approach significantly reduces environmental impact and creates new economic opportunities. According to a study by the Fraunhofer Institute, efficient battery recycling can recover up to 95% of valuable materials such as lithium, cobalt and nickel. In addition, companies specializing in battery recycling are emerging, promoting the development of advanced technologies and generating employment in sustainable sectors.

Conclusion

The environmental benefits of electric vehicles are numerous and varied. According to a report by the International Energy Agency (IEA), electric cars significantly reduce greenhouse gas emissions, contributing to climate change mitigation. A study from the University of Michigan found that electric vehicles can reduce CO_2 emissions by 50% compared to internal combustion vehicles, even when the electricity comes from a mix of energy sources.

In addition, electric vehicles improve air quality by not emitting pollutants such as nitrogen oxides (NOx) and particulate matter (PM), which are associated with respiratory and cardiovascular problems. A report by the World Health Organization (WHO) highlights that reducing these pollutants can reduce the incidence of respiratory diseases and improve public health.

The adoption of electric vehicles also encourages the conservation of natural resources. By reducing the demand for oil, the need for fossil fuel exploration and extraction is reduced, protecting sensitive ecosystems and reducing the risk of spills and environmental disasters. In addition, innovations in battery technologies are leading to the development of more sustainable materials and the efficient recycling of components, promoting a circular economy.

As we move towards a more environmentally conscious society, it is crucial to support and accelerate the transition to electric vehicles. A global commitment and concerted effort can harness the environmental advantages of electric vehicles to combat climate change, improve public health, and preserve our planet for future generations. According to the Intergovernmental Panel on Climate Change (IPCC), the transition to electric mobility is a key strategy to limit global warming to 1.5°C and ensure a sustainable future.

Current challenges: Infrastructure, autonomy and cost

Despite the numerous environmental and operational advantages of electric vehicles (EVs), their mass adoption faces a number of significant challenges. These challenges must be addressed to facilitate a faster and more efficient transition to more sustainable mobility.

One of the main obstacles is the charging infrastructure. The lack of a network of widely available fast-charging stations can lead to anxiety in drivers about the range of their vehicles, known as "range anxiety." Although efforts are underway to expand infrastructure, there is still much to be done to make EV charging as convenient as refueling.

Battery range is another crucial challenge. Although battery technologies have advanced significantly, many electric vehicles still can't match the range of internal combustion vehicles on a single tank of gas. Research efforts are focused on improving energy density and reducing charging times, but these advances have not yet become widespread.

Finally, the cost of acquiring and maintaining electric vehicles remains a barrier. Although prices have declined over time and are expected to continue to fall, electric vehicles are typically more expensive than their internal combustion equivalents due to the high cost of batteries. Additionally, consumers may be concerned about the cost and availability of specialized maintenance services. To overcome these challenges, supportive policies, financial incentives, and a continued focus on technological innovation are needed.

1. Charging Infrastructure

Charging infrastructure is critical to the widespread adoption of electric vehicles. Without a proper charging network, users may experience "range anxiety," the fear of running out of charge before reaching their destination.

Types of Charging Stations

Home Charging

Pros: Home charging is a convenient option for EV owners, allowing them to recharge their vehicles overnight. Installing a home charger makes daily charging easier and reduces reliance on public stations, offering convenience and time savings.

Challenges: Not every homeowner has access to a garage or private parking spot where they can install a charger. In addition, the charging capacity of domestic electrical installations can be limited, especially in older homes.

Types: Public charging stations are divided into three main categories:

- o **Level 1**: Uses a standard 120V power outlet and provides slow charging, ideal for overnight charging at home.
- o **Level 2**: Uses a 240V power outlet and is common in public charging stations, offering faster charging and suitable for commercial and public parking lots.
- o **Level 3 (DCFC):** Provides ultra-fast charging using direct current, allowing electric vehicles to recharge the battery up to 80% in about 30 minutes. These stations are ideal for long journeys and are located on highways and high-demand corridors.

Challenges: The distribution and availability of public charging stations are uneven, with a higher concentration in urban areas and less coverage in rural areas. A lack of standardization in connectors and charging protocols can also be a hurdle, complicating access and use for EV drivers.

Infrastructure Development and Expansion

Government Initiatives

Public Investments: Many governments are investing in the expansion of public charging infrastructure. Grant and funding programs are aimed at increasing the number of charging stations on highways, urban areas, and shopping malls.

Policies and Regulations: Some governments are implementing regulations that require the installation of charging stations in new residential and commercial developments, as well as in public buildings and parking lots. These policies encourage a more accessible and convenient environment for EV owners.

Private Sector Engagement

Energy and Automotive Companies: Energy companies, automakers, and technology companies are investing in the creation of charging networks. For example, Tesla has developed its network of superchargers, while companies like ChargePoint and EVgo are expanding their public charging networks.

Collaborations and Partnerships: Collaborations between private sector companies and public entities are facilitating the expansion of charging infrastructure, combining resources and knowledge to implement effective solutions. These partnerships are essential to developing a robust and accessible charging network.

Wireless Charging: Wireless charging, which allows electric vehicles to be recharged without needing to be plugged into a socket, is being developed and tested. This technology can simplify charging and increase convenience for users, eliminating the need for cables.

Bidirectional charging: Bidirectional charging allows electric vehicles to return energy to the power grid (vehicle-to-grid, or V2G) or power devices and homes (vehicle-to-home, or V2H). Not only does this improve energy management, but it can also provide additional revenue to EV owners, creating a more efficient and flexible energy system.

2. Battery Autonomy

Battery range is another crucial challenge for electric vehicles. "Range anxiety" is a common concern among potential buyers, who fear running out of charge while traveling.

Advances in Battery Technology

Lithium-Ion Batteries

Lithium-ion batteries are currently the predominant technology in electric vehicles. They stand out for their good energy density and reasonable charging times. However, they still face limitations in terms of autonomy and cost. Thanks to continuous advances in battery chemistry, thermal management, and control systems, the capacity and durability of these batteries are improving, resulting in greater range for electric vehicles.

Solid State Batteries

Solid-state batteries represent a significant innovation by using solid electrolytes instead of liquids. This can significantly increase energy density, reduce charging times, and improve battery safety. Although promising, these batteries are still in the research and development phases. They currently face technical and production challenges that must be overcome before they can be commercialized on a large scale.

Other Battery Technologies

Graphene Batteries

Graphene-based batteries have the potential to offer significant improvements in storage capacity and charging speed. However, this technology still requires more research and development before it is ready for commercialization.

Metal-Air Batteries

Metal-air batteries, which use metals such as lithium or zinc and oxygen from the air to generate energy, have a high energy density. However, they face challenges in terms of durability and life cycle. This technology promises great advances, but it also needs to overcome technical hurdles before it can be widely adopted in the market.

Strategies to Increase Autonomy

Vehicle Design Optimization

Improving the aerodynamics of electric vehicles is crucial to reduce air resistance and increase energy efficiency, which in turn extends range. In addition, the use of lightweight yet strong materials, such as carbon fiber and aluminum alloys, can significantly reduce the weight of the vehicle, further improving its energy efficiency.

Energy Management Systems

Regenerative braking systems are a key innovation in electric vehicles. These systems capture kinetic energy during braking and convert it into electricity to recharge the battery, which helps to increase the vehicle's range. In addition, many electric vehicles offer driving modes that optimize energy use by adjusting acceleration and other parameters, thus maximizing the vehicle's range and efficiency.

3. Acquisition and Maintenance Cost

The cost of acquisition and maintenance is one of the main obstacles to the mass adoption of electric vehicles. Although costs are falling, EVs are still typically more expensive than their internal combustion counterparts.

Acquisition Cost

The price of batteries is a crucial factor in the total cost of an electric vehicle. Although lithium-ion battery prices have dropped dramatically over the past decade, they still account for a significant portion of the cost of the vehicle. However, costs are expected to continue to decline as production increases and manufacturing technologies improve, thanks to economies of scale.

Government subsidies and incentives play an important role in the adoption of electric vehicles. Many governments offer tax breaks, subsidies, and tax breaks that can significantly reduce the initial cost of these vehicles. In addition, leasing programs and accessible financing options make electric vehicles more affordable for consumers, incentivizing their mass adoption.

Maintenance Cost

Electric vehicles have fewer moving mechanical components compared to internal combustion vehicles. This results in fewer parts that can wear out or need replacement, thus reducing maintenance costs. In addition, regenerative braking not only helps to increase range, but also reduces brake wear, decreasing the need for frequent replacements of brake pads and discs.

The durability of the batteries is another significant advantage. Many EV manufacturers offer extended warranties for batteries, covering 8 to 10 years or up to 100,000 miles, giving consumers peace of mind and reducing the potential cost of battery replacement. In addition, used EV batteries can be recycled or reused in energy storage applications, which can reduce costs and environmental impact.

4. Additional Challenges and Potential Solutions

Consumer Awareness and Acceptance

In addition to charging infrastructure, range, and cost, there are other challenges that electric vehicles face today. A crucial aspect is consumer awareness and acceptance.

Education and Information: Educational campaigns can help inform consumers about the benefits of electric vehicles and dispel myths and misunderstandings. Offering free test drives or demonstration events allows consumers to experience first-hand the advantages of these vehicles.

Non-Financial Incentives: Allowing electric vehicles to use high-occupancy (HOV) lanes or bus-only lanes can increase their appeal. In addition, providing preferential parking for electric vehicles in urban areas can incentivize their adoption.

Environmental Impact of Battery Production

The environmental impact of battery production is another major challenge.

Sustainability in the Supply Chain: Adopting responsible and ethical mining practices for the extraction of materials such as lithium, cobalt, and nickel is crucial to minimizing environmental and social impact. Fostering a circular economy through recycling and reusing materials can reduce demand for new raw materials and decrease environmental impact.

Innovation in Materials and Processes: Researching and developing new battery materials that are more abundant and less harmful to the environment can help reduce the impact of battery production. In addition, optimizing manufacturing processes to reduce energy consumption and emissions can make battery production more sustainable.

Conclusion

Electric vehicles (EVs) offer significant promise for a cleaner and more sustainable transport future. However, its mass adoption faces significant challenges related to charging infrastructure, battery autonomy, and acquisition and maintenance costs. Addressing these challenges requires a multifaceted approach that combines government efforts, technological innovation, and public-private sector collaboration.

Charging infrastructure is critical to the viability of EVs. Although progress is being made in expanding charging stations, it is crucial to increase their availability in urban and rural areas to reduce range anxiety. In addition, the continuous improvement in energy density and cost reduction of lithium-ion batteries, as well as the development of new technologies such as solid-state batteries, can extend the range and make EVs more competitive in the market.

Acquisition and maintenance costs are also a significant challenge. Although battery prices have declined, they are still a considerable part of the total cost of EVs. Government subsidies and incentives, such as tax breaks and leasing programs, can help make EVs more accessible to consumers. In addition, EVs have fewer moving mechanical components, reducing maintenance costs in the long run.

As EV technologies continue to advance and supporting infrastructures expand, these challenges are likely to be mitigated and EVs will become an increasingly viable and attractive option for consumers. With global commitment and concerted action, EVs have the potential to transform the transport industry and contribute significantly to the fight against climate change, improving air quality and promoting a more sustainable future for all.

Chapter 5: Innovation and the Future: The Toyota Case

Toyota: Pioneers in electrification

Toyota, one of the world's most recognizable automotive brands, has been a pioneer in the adoption of electrification technologies. From the launch of the first mass-produced hybrid vehicle to its current focus on electric and hydrogen vehicles, Toyota has played a crucial role in transforming the automotive industry towards a more sustainable future.

First Steps Toward Electrification: The Toyota Prius

Toyota's history in electrification begins with the launch of the Prius in Japan in 1997 and in the global market in 2000. This hybrid vehicle combined an internal combustion engine with an electric motor and a rechargeable battery, using electric power for low-speed driving and gasoline for power and range. Not only did the Prius improve fuel efficiency through kinetic energy recovery during braking, but it also set a new standard in the automotive industry. The Synergy Drive Hybrid system enabled the efficient joint operation of the gasoline engine and electric motor, contributing to its commercial success and the adoption of hybrid technologies by other manufacturers.

Expansion and Diversification: Beyond the Prius

Following the success of the Prius, Toyota expanded its lineup of hybrid vehicles to include models such as the Camry Hybrid, Highlander Hybrid, and RAV4 Hybrid, each tailored to different consumer needs and preferences. The Camry Hybrid, launched in 2006, combined the popular design of the Camry sedan with hybrid technology, offering a more efficient and environmentally friendly option. The Highlander Hybrid, an SUV launched in 2005, provided an eco-friendly alternative in the sport utility vehicle segment, combining space, power and efficiency. In 2016, Toyota introduced the RAV4 Hybrid, a hybrid version of the popular compact SUV that offered greater fuel efficiency without compromising performance and versatility.

In addition, Toyota introduced the Prius Prime in 2012, a plug-in version of the Prius that allowed drivers to recharge the battery via an electrical outlet, offering greater range in electric mode. This model represented a significant advancement in hybrid technology, providing consumers with an even more sustainable and efficient option.

Global Electrification Strategy

Toyota has developed a comprehensive strategy that encompasses hybrid, electric and hydrogen vehicles. In 2021, it introduced the bZ4X, its first mass-produced battery electric vehicle, built on the e-TNGA platform designed specifically for electric vehicles. This electric SUV marks the beginning of a new era for Toyota in the electric vehicle market. The e-TNGA platform allows for great flexibility in the design and development of future electric models, facilitating the expansion of Toyota's electric vehicle range.

In 2014, Toyota launched the Mirai, one of the first commercially available hydrogen fuel cell vehicles. The Mirai uses hydrogen to generate electricity through a fuel cell, emitting only water vapor as a byproduct. Toyota has also been an active advocate for hydrogen infrastructure development, collaborating with governments and businesses to establish hydrogen fueling stations around the world.

Toyota Technology Innovations

Toyota has been a leader in technological innovation, developing advanced energy management systems such as the Synergy Drive Hybrid and perfecting regenerative braking technology. The Synergy Drive Hybrid system has evolved significantly since its introduction in Prius, with improvements in energy management, electric motor efficiency and battery capacity. These advancements have enabled Toyota's hybrid vehicles to deliver superior performance and efficiency, applying to a wide range of vehicles, from compact to SUVs and commercial vehicles.

Toyota has made significant investments in the research and development of lithium-ion batteries and solid-state batteries. Improvements in energy density, safety, and charging times have allowed Toyota to offer electric vehicles with greater range and performance. The company is at the forefront of research into solid-state batteries, a technology that promises to revolutionize the automotive industry by offering higher energy density, faster charging times, and greater safety compared to lithium-ion batteries.

In the area of autonomous driving and connectivity, Toyota has developed the Toyota Safety Sense system, which includes advanced driver assistance technologies such as automatic emergency braking, adaptive cruise control, lane departure warning and traffic sign recognition. These technologies improve driving safety and comfort. Toyota continues to improve and update its ADAS systems, incorporating new features and capabilities to stay at the forefront of driver assistance technology.

Commitment to Sustainability and Social Responsibility

Toyota has demonstrated a strong commitment to sustainability and social responsibility. In 2015, it launched the Toyota 2050 Environmental Challenge, an ambitious initiative that sets six key goals to reduce the company's environmental impact. These goals include eliminating CO_2 emissions from new vehicles and factories, reducing water use, and promoting the circular economy. Toyota has made significant progress towards these goals, including improving the energy efficiency of its vehicles, implementing more sustainable production practices, and expanding its range of electrified vehicles.

The company is increasing the use of renewable energy in its global operations, including installing solar panels and purchasing electricity from renewable sources. Toyota is also adopting technologies and processes that reduce CO_2 emissions across all its activities, from vehicle production to transportation and logistics.

Future of Electrification at Toyota

Toyota continues to lead the way in electrification with plans to launch new battery electric vehicle (BEV) models and improve its hybrid and hydrogen technologies. The company is collaborating with governments and industrial partners to expand charging infrastructure and develop fast-charging technologies. Toyota is at the forefront of solid-state battery research, with plans to commercialize this technology in the next decade. These batteries offer higher energy density, increased safety, and faster charging times compared to current lithium-ion batteries.

Toyota is also developing advanced battery recycling and reuse methods to minimize environmental impact and promote a circular economy in the automotive industry. The company is integrating smart mobility technologies, such as autonomous driving and connectivity, to improve safety, efficiency, and user experience. In addition, Toyota is exploring Mobility as a Service (MaaS) solutions, providing ride-sharing options and integrated mobility services that reduce the need for vehicle ownership and promote efficient use of resources.

Conclusion

Toyota has been a pioneer in electrification and sustainability, leading the way with technological innovations and a strong commitment to the environment. From the launch of the Prius to the development of electric and hydrogen vehicles, Toyota has demonstrated its ability to adapt and lead in an ever-evolving market. With a clear vision and a comprehensive strategy, Toyota is well positioned to continue leading the transformation towards cleaner and more efficient transportation, contributing significantly to the fight against climate change and the protection of the planet for future generations.

Toyota: Pioneer in Technological Innovation and Sustainable Mobility

Toyota, one of the largest and most recognized automotive companies in the world, has been an undisputed leader in technological innovation. Throughout its history, Toyota has developed and perfected numerous technologies that have revolutionized the automotive industry. From advanced hybrid systems to autonomous driving and hydrogen fuel cells, Toyota's innovations have paved the way to a cleaner, safer and more efficient future.

Toyota Technology Innovations

Hybrid Technology: The Heart of Toyota's Revolution

Toyota's Synergy Drive Hybrid (HSD) is one of the company's most iconic technologies. First introduced with the Toyota Prius in 1997, the HSD combines an internal combustion engine with an electric motor and rechargeable battery, offering unprecedented fuel efficiency and emissions reduction.

The HSD system uses both engines optimally. At low speeds and in light driving conditions, the electric motor drives the vehicle, using the energy stored in the battery. As speed increases or more power is required, the internal combustion engine kicks in, working in tandem with the electric motor. In addition, the HSD system incorporates regenerative braking, which recovers kinetic energy during braking and converts it into electricity to recharge the battery. Not only does this improve fuel efficiency, but it also reduces brake wear.

Successive generations of the HSD have seen significant improvements in terms of efficiency, performance and weight reduction. Toyota has optimized the components of the hybrid system, including more efficient electric motors, higher-capacity batteries and more advanced energy management systems. The HSD has not been limited to the Prius, but has been integrated into a wide range of models, from sedans like the Camry Hybrid to SUVs like the RAV4 Hybrid and Highlander Hybrid.

Plug-in Hybrids (PHEV)

Toyota has also pioneered the development of plug-in hybrid vehicles, which combine the advantages of conventional hybrids with the ability to recharge the battery via a power outlet. Launched in 2012, Prius Prime offers longer range in all-electric mode thanks to a larger battery capacity. This model allows drivers to make most of their daily journeys without consuming petrol. Toyota expanded its lineup of plug-in hybrids with the launch of the RAV4 Prime, an SUV that combines impressive performance with significant electric range.

Hydrogen Fuel Cell Vehicles (FCEVs)

Toyota has been a fervent advocate of hydrogen fuel cells as a viable and sustainable solution for future mobility. Fuel cell vehicles generate electricity through a chemical reaction between hydrogen and oxygen, emitting only water vapor as a byproduct. The Toyota Mirai, first launched in 2014, is one of the first commercially available hydrogen fuel cell vehicles. This vehicle uses hydrogen stored in high-pressure tanks and oxygen from the air to generate electricity through a fuel cell, powering the vehicle's electric motor and emitting only water vapor.

The second generation of the Mirai, launched in 2020, features significant improvements in design, performance and range. Toyota has actively worked to promote and expand hydrogen infrastructure, collaborating with governments and industrial partners to establish hydrogen fueling stations around the world.

Battery Electric Vehicles (BEVs)

In response to the growing demand for electric vehicles and the need to reduce carbon emissions, Toyota has stepped up its efforts in the development of battery electric vehicles (BEVs). Toyota has developed the e-TNGA (Toyota New Global Architecture) platform specifically for battery electric vehicles. This modular and flexible platform enables the efficient development of a wide range of electric models.

The e-TNGA platform can adapt to different battery sizes, drive configurations (front, rear, or all-wheel-drive), and body types, allowing

Toyota to design and produce various electric models efficiently. The bZ4X, launched in 2021, is Toyota's first mass-produced battery electric vehicle based on this platform. This electric SUV offers competitive range and fast-charging options, as well as being equipped with the latest driver assistance, connectivity and entertainment technologies.

Autonomous Driving and Connectivity

Toyota is investing significantly in autonomous driving and connectivity technologies to improve safety, efficiency, and user experience. Toyota Safety Sense (TSS) is a set of advanced driver assistance technologies designed to prevent accidents and protect vehicle occupants. It includes features such as automatic emergency braking, adaptive cruise control and lane departure warning.

Toyota is developing autonomous driving technologies through its Toyota Research Institute (TRI) division and collaborations with technology companies. Level 2 systems offer advanced driver assistance, including adaptive cruise control, lane centering and traffic jam assist. Toyota is also researching Level 4 and 5 technologies, which will enable fully autonomous driving without human intervention. The company is developing Woven City, a prototype city in Japan that will serve as a living laboratory for testing advanced mobility technologies, including autonomous vehicles, robotics and smart energy.

Sustainability and Renewable Energy

Toyota's commitment to sustainability extends beyond vehicle electrification, encompassing renewable energy production and green practices throughout the supply chain. The Toyota 2050 Environmental Challenge is an ambitious initiative that sets six key goals to reduce the company's environmental impact. These goals include reducing CO_2 in new vehicles, eliminating CO_2 emissions in factories, and minimizing water use. Toyota promotes recycling and reuse of materials throughout its supply chain to minimize waste and conserve natural resources.

Toyota has significantly increased the use of renewable energy in its global operations, including the installation of solar panels and the purchase of electricity from renewable sources. The company has implemented water conservation and recycling programs in its factories, achieving substantial reductions in resource use and waste generation. Toyota is also adopting sustainable production practices, such as Lean and Green manufacturing, which maximizes efficiency and minimizes waste, and the use of clean technologies in its factories.

Conclusion

Toyota has been a leader in technological innovation in the automotive industry, developing and perfecting technologies that have revolutionized sustainable mobility. From the Synergy Drive Hybrid to hydrogen fuel cell vehicles and BEVs, Toyota's innovations have set new standards in efficiency, performance and sustainability.

Toyota's commitment to research and development, sustainability, and social responsibility has allowed the company to remain at the forefront of the automotive industry. As the company continues to move towards a cleaner and more sustainable future, its technological innovations will continue to play a crucial role in transforming mobility and protecting the environment. With a clear vision and a comprehensive strategy, Toyota is well-positioned to meet the challenges of the future and lead the transition to cleaner, safer and more efficient transportation. Toyota's technological innovations have not only transformed the automotive industry, but have also demonstrated the power of innovation to create a more sustainable and better future for all.

The Impact of Toyota's Electric Vehicles on the Global Market

Toyota, one of the world's most influential automotive companies, has played a crucial role in the evolution of electric vehicles (EVs) and the transformation of the global market towards more sustainable mobility. From the launch of the pioneering Prius to its advanced battery electric vehicle (BEV) and hydrogen fuel cell (FCEV) models, Toyota has demonstrated an ongoing commitment to innovation and sustainability. This section explores in depth the impact of Toyota's electric vehicles on the global market, highlighting their technological contributions, changes in consumer behavior, and the implications for the automotive industry as a whole.

The Toyota Prius: Starting the Electric Vehicle Revolution

The launch of the Toyota Prius in 1997 marked a significant milestone in the history of electric vehicles. It was the first mass-produced hybrid vehicle and laid the foundation for the global acceptance of hybrid technology. Prius introduced the Synergy Drive Hybrid (HSD) system, a combination of an internal combustion engine and an electric motor that offered unprecedented fuel efficiency and reduced greenhouse gas emissions. The Prius' efficiency and ability to operate in all-electric mode at low speeds demonstrated that hybrid vehicles could be a viable alternative to traditional internal combustion vehicles.

The Prius was well received in global markets, especially in Japan and the United States. Its commercial success helped change consumers' perception of electric vehicles, highlighting their environmental and economic benefits. In addition, the Prius received numerous awards and recognitions, including the prestigious "Car of the Year" in several countries, cementing its reputation as an innovative and sustainable vehicle.

The success of the Prius influenced the implementation of stricter emission standards in several countries. Environmental regulations began to favor clean and efficient technologies, prompting other manufacturers to develop hybrid and electric vehicles. The Prius became the gold standard for other manufacturers looking to develop hybrid technologies. Many companies began investing in research and development to create their own hybrid models, inspired by the success of the Prius.

Toyota expanded its lineup of hybrid vehicles, launching models such as the Camry Hybrid, Highlander Hybrid, and RAV4 Hybrid. This diversification allowed Toyota to capture a larger market share and meet diverse consumer needs and preferences. The success of the Prius fostered competition and collaboration in the automotive industry. Other manufacturers launched their own hybrid models, and some companies established strategic alliances with Toyota to leverage its expertise in hybrid technology.

Expansion to Battery Electric Vehicles (BEVs)

In response to the growing demand for battery electric vehicles, Toyota developed the e-TNGA platform, a modular architecture designed specifically for BEVs. The launch of the bZ4X in 2021 was an important milestone in Toyota's electrification strategy. The e-TNGA platform is highly flexible, allowing the production of a variety of BEV models with different battery sizes, drive configurations and body types. This adaptability facilitates the rapid expansion of Toyota's EV lineup.

The bZ4X offers competitive autonomy and robust performance, suitable for a wide range of uses. The battery capacity and efficiency of the electric propulsion system ensure a reliable and long-lasting driving experience. In addition, the bZ4X is equipped with the latest technologies in driver assistance, connectivity and entertainment, providing a modern and safe driving experience.

Toyota plans to launch several additional models based on the e-TNGA platform in the coming years, spanning different market segments. This expansion will allow Toyota to meet the growing demand for BEVs and compete effectively in the global market. The expansion of Toyota's BEV lineup reflects its commitment to sustainability and reducing carbon emissions. By offering a variety of electric options, Toyota is making a significant contribution to the transition to cleaner and more efficient transportation.

Innovations in Hydrogen Fuel Cell Vehicles (FCEVs)

The Toyota Mirai, first launched in 2014, is one of the first commercially available hydrogen fuel cell vehicles. It represents a significant advancement in hydrogen technology and Toyota's vision for an emission-free future. The Mirai's fuel cell system uses hydrogen stored in high-pressure tanks and oxygen from the air to generate electricity through a chemical reaction in the fuel cell. This electricity powers the vehicle's electric motor, and the Mirai's only emission is water vapor, making it one of the cleanest vehicles available on the market.

The second generation of the Mirai, launched in 2020, features significant improvements in terms of design, performance and range. The new Mirai offers a more refined and engaging driving experience, with a longer range and a fast hydrogen refueling time. Toyota has actively worked to promote and expand hydrogen infrastructure, collaborating with governments and industrial partners to establish hydrogen fueling stations around the world.

Toyota has been a key advocate in the development of hydrogen infrastructure, collaborating with governments and companies to establish hydrogen refueling stations in key markets. This infrastructure is essential for the widespread adoption of FCEVs. Toyota has conducted education and awareness campaigns to inform the public about the benefits of hydrogen as a clean and sustainable energy source. These efforts help build buy-in and support for FCEVs.

Toyota continues to invest in the research and development of hydrogen fuel cell technologies, seeking to improve the efficiency, reduce costs, and increase the durability of fuel cell systems. Toyota plans to expand its line of fuel cell vehicles with new models in different market segments. This expansion will help meet the growing demand for clean and sustainable transportation solutions.

Impact on the Automotive Industry and Consumer Behavior

Toyota's electric vehicles have had a profound impact on the automotive industry, driving innovation and competitiveness in the global market. Toyota has set new standards in electric, hybrid and fuel cell vehicle technology, prompting other manufacturers to invest in research and development to stay competitive. Competition and collaboration between car manufacturers have accelerated the development of new technologies and the introduction of more efficient and sustainable models to the market.

The success of Toyota's electric vehicles has influenced the implementation of stricter emission standards in many countries, promoting the adoption of clean technologies and the reduction of carbon emissions. Governments around the world have introduced financial incentives and favorable policies to encourage the adoption of electric vehicles, which has further boosted demand and the development of supporting infrastructure.

Toyota's electric vehicles have played a crucial role in changing consumer perceptions and behaviors towards sustainable mobility. Toyota has conducted education and awareness campaigns to inform consumers about the environmental and economic benefits of electric vehicles. These efforts have increased public awareness of the need to adopt more sustainable transport solutions. As consumers become more aware of the environmental impacts of their purchasing decisions, there has been an increase in the preference for electric and hybrid vehicles. Toyota has capitalized on this trend by offering a wide range of sustainable options.

Toyota's electric vehicles are equipped with advanced connectivity and driver assistance technologies, enhancing the user experience and making the transition to electric mobility more attractive. Electric vehicles offer significant savings in fuel and maintenance costs, appealing to consumers looking for economical and efficient options.

Future of Toyota's Electric Vehicles in the Global Market

Toyota is well-positioned to continue to lead the EV market in the future, with clear strategies for the expansion and continued development of clean technologies. Toyota plans to launch a number of new BEV and FCEV models in the coming years, covering different market segments and meeting various consumer needs. The company will continue to invest in the research and development of new battery and fuel cell technologies, improving the efficiency, range, and accessibility of its electric vehicles.

Toyota will work collaboratively with governments and industry partners to expand EV charging infrastructure, ensuring consumers have access to fast and convenient charging stations. The company will also continue to promote the development of hydrogen fueling stations, facilitating the adoption of FCEVs and supporting the transition to a hydrogen economy.

Toyota has played a pivotal role in transforming the automotive industry towards a more sustainable future, through its innovations in electric, hybrid and fuel cell vehicles. The impact of Toyota's electric vehicles on the global market is undeniable, driving innovation, influencing regulation and shifting consumer behaviour towards cleaner and more efficient mobility.

As Toyota continues to develop new technologies and expand its electric vehicle offerings, it is well-positioned to lead the transition to sustainable transportation and play a crucial role in the fight against climate change. With an ongoing commitment to sustainability, innovation and excellence, Toyota continues to lead the way to a cleaner, safer and more efficient future for all.

Chapter 6: Global Policies for a Sustainable Future

Government initiatives and incentive policies

In a world facing the growing urgency of combating climate change and protecting the environment, governments around the world are implementing a variety of incentive initiatives and policies to promote a sustainable future. These measures range from financial incentives and strict regulations to research and development programs and international agreements. In this section, we will explore in depth the government initiatives and incentive policies that are driving the transition to a greener and more sustainable economy.

Financial Incentives for Clean Technology Adoption

Tax Subsidies and Discounts

One of the most effective methods that governments have used to encourage the adoption of clean technologies is through financial incentives. Direct subsidies and tax rebates are two key tools in this strategy. For example, in Norway and Germany, buyers of electric vehicles (EVs) can receive subsidies of several thousand dollars, which has led to increased adoption of these vehicles. Additionally, in countries like the United States, tax incentives allow EV buyers to qualify for a federal tax credit of up to $7,500. These incentives are not limited to vehicles; Governments also offer subsidies for the installation of renewable energy systems, such as solar panels and wind turbines, reducing upfront costs and encouraging investment in clean energy technologies.

Financing and Leasing Programs

In addition to subsidies and tax rebates, many governments have implemented financing and leasing programs to facilitate EV procurement. Financing programs with low-interest loans and favorable leasing terms make EVs more accessible to consumers and businesses. Governments also create green investment funds to finance renewable energy and energy efficiency projects, providing capital for the development of new clean energy facilities and the modernization of existing infrastructure.

Strict Environmental Regulations and Standards

Emissions and Energy Efficiency Standards

Government regulations play a crucial role in promoting clean technologies. The European Union, for example, has implemented the Euro Standards, which set strict limits for emissions of CO_2 and other pollutants from new vehicles. In the United States, the Environmental Protection Agency (EPA) sets emission standards for vehicles through the Corporate Average Fuel Economy (CAFE) program. These rules force automakers to develop cleaner technologies to meet increasingly stringent standards.

Prohibitions and Restrictions

In addition to emissions standards, some governments have announced plans to ban the sale of new internal combustion vehicles in the near future. Norway, for example, plans to ban the sale of new petrol and diesel cars by 2025. Other countries, such as the United Kingdom and France, have set deadlines for 2030 and 2040, respectively. These bans encourage the adoption of electric vehicles and other clean modes of transportation.

Research and Development (R+D) Programs

Investments in technological innovation

Research and development (R+D) is essential to advance clean technologies. Governments are investing in solar and wind energy projects, as well as the development of energy storage technologies. These investments not only improve the efficiency and reduce the costs of existing technologies, but also drive the creation of new sustainable solutions. In addition, research into electric and hydrogen vehicles, as well as the development of charging infrastructure, are key areas of investment that help improve the range and reduce the costs of sustainable vehicles.

Public-Private Partnerships

Public-private partnerships are key to accelerating technological innovation. Innovation consortia bring together companies, universities, and government organizations to work on research and development projects. These alliances facilitate the exchange of knowledge and resources, accelerating the development of new sustainable technologies. Smart city initiatives are another example of public-private collaboration, integrating sustainable mobility, energy management, and smart utility technologies to create more efficient and sustainable urban environments.

International Agreements and Global Cooperation

International Climate Agreements

The Paris Agreement, adopted in 2015, is one of the most important international treaties in the fight against climate change. It aims to limit global temperature rise to less than 2 degrees Celsius above pre-industrial levels, with efforts to limit it to 1.5 degrees. The signatory countries commit to reducing their greenhouse gas emissions and to submit national climate action plans (NDCs). The Kyoto Protocol, adopted in 1997, was the precursor to the Paris Agreement and set emission reduction targets for developed countries.

Cooperation in Energy and Technology

Organizations such as the International Renewable Energy Agency (IRENA) promote the adoption and sustainable use of renewable energy through advice, technical support, and financing for renewable energy projects around the world. International cooperation in research and development, as well as the exchange of good practices, are essential to accelerate the transition to clean and sustainable technologies.

Impact of Incentive Initiatives and Policies

Positive Results and Benefits

Government initiatives and incentive policies have contributed significantly to the reduction of greenhouse gas emissions, improving air quality and fostering the growth of the renewable energy industry. Installed solar and wind power capacity has increased significantly, reducing reliance on fossil fuels and diversifying energy sources. In addition, the transition to a green economy has generated new jobs in sectors such as renewable energy, energy efficiency and sustainable mobility, contributing to economic development.

Challenges and Opportunities

Despite progress, financing and resource allocation remain challenges for the implementation of sustainable initiatives, especially in developing countries. The adoption of new technologies and sustainable practices may face resistance due to lack of knowledge, perception of high costs, or institutional inertia. However, continued innovation in clean and sustainable technologies offers opportunities to improve efficiency, reduce costs, and increase the adoption of green solutions. International cooperation is essential to address the global challenges of climate change and sustainability, enabling countries to share knowledge, resources and technologies to ensure a sustainable future for all.

Conclusion

Government initiatives and incentive policies are critical to driving the transition to a sustainable future. Through financial incentives, strict regulations, research and development programs, and international agreements, governments are promoting the adoption of clean technologies, emission reductions, and environmental protection. Although there are challenges in implementing and adopting these policies, the opportunities for innovation and international cooperation are immense. With continued commitment and concerted action, it is possible to achieve a cleaner, safer and more sustainable future for future generations. The policies and measures described in this section are just the beginning; The path to sustainability requires constant evolution and a collective global effort.

Emission standards and their impact

In the fight against climate change, emissions standards have become crucial tools for reducing the amount of greenhouse gases and other pollutants released into the atmosphere. These regulations not only seek to mitigate environmental impact, but also to protect public health and encourage technological innovation. In this chapter, we will explore in detail the various emissions standards implemented globally, their effectiveness, and their impact on different sectors.

Evolution of Emission Standards

Early Initiatives and International Agreements

Kyoto Protocol (1997)

The Kyoto Protocol was one of the first international agreements to commit developed countries to reducing their greenhouse gas emissions. It set binding targets and promoted flexibility mechanisms such as emissions trading and reduction projects in developing countries. Although limited in scope, it laid the groundwork for future negotiations and highlighted the need for a coordinated global effort to address climate change.

Paris Agreement (2015)

The Paris Agreement seeks to limit global temperature rise to less than 2 degrees Celsius above pre-industrial levels, with efforts to limit it to 1.5 degrees. All countries must submit national climate action plans (NDCs) and review their commitments regularly. This agreement has fostered greater climate ambition, with many countries implementing stricter policies and increasing their emission reduction commitments.

Regional and National Standards

European Union Emission Standards (Euro)

The European Union has implemented a series of standards known as the Euro Standards, which set strict limits for CO_2, NOx and particulate emissions from new vehicles. These standards have become progressively stricter since their introduction in 1992, forcing automakers to develop cleaner technologies, such as more efficient engines and exhaust after-treatment systems. They have also accelerated the adoption of electric and hybrid vehicles.

United States Emission Standards (CAFE)

The Corporate Average Fuel Economy (CAFE) program establishes fuel efficiency requirements for vehicles sold in the United States. Introduced in the 1970s, it has been regularly updated to reflect current environmental and technological needs. CAFE standards have led to significant improvements in vehicle fuel efficiency, reducing greenhouse gas emissions and dependence on oil.

China Emission Standards

In response to severe air pollution, China has implemented emission standards similar to the Euro Standards. The China VI Standards, which came into force in 2020, are some of the strictest in the world and cover both light and heavy-duty vehicles. These regulations have led to a noticeable decrease in pollution levels in many Chinese cities, improving public health and reducing the incidence of respiratory diseases.

Impact on the Automotive Sector

Technological Innovations

Electric and Hybrid Vehicle Development

Emissions standards have been a key catalyst for the development and adoption of electric and hybrid vehicles. These vehicles produce significantly fewer emissions than traditional internal combustion vehicles. The need to meet emissions standards has driven innovations in battery technology, electric motors, and energy management systems. These improvements have increased range, reduced charging times, and improved the overall efficiency of electric vehicles.

Exhaust Gas Aftertreatment Systems

To meet strict emission limits, automakers have developed advanced aftertreatment systems such as particulate filters and three-way catalytic converters. These systems remove or significantly reduce pollutants before they are released into the atmosphere. The widespread implementation of these systems has led to substantial improvements in air quality, especially in urban areas with high traffic density.

Challenges and Opportunities

Compliance Costs

Developing clean technologies and adapting production lines to meet emissions standards requires significant investments in research and development. Although this can increase upfront costs, it also creates opportunities for innovation and differentiation in the market. As more manufacturers adopt clean technologies, production costs tend to decrease thanks to economies of scale and competition in the component supplier market.

Market Adoption

The adoption of cleaner vehicles also depends on consumer education and awareness. It is important for consumers to understand the environmental and economic benefits of low-emission vehicles. Financial incentives, such as subsidies and tax breaks, can help accelerate the adoption of clean technologies by reducing the upfront cost for consumers.

Impact on other industrial sectors

Many countries have implemented strict standards to limit CO_2 and NOx emissions from power plants. These regulations encourage the use of cleaner technologies, such as carbon capture and storage (CCS) and conversion to natural gas. Integrating renewable energy sources into the electricity grid requires greater flexibility and decentralization. Energy storage technologies, such as large-scale batteries, and smart grids are essential to manage intermittency and ensure a stable electricity supply.

Emission standards also apply to industrial processes, including the production of steel, cement and chemicals. These regulations require the use of emission control technologies, such as filters and scrubbers. The need to meet emissions standards has led to innovations in manufacturing processes, including the use of cleaner, more efficient materials and the optimization of production processes to reduce emissions. Emissions policies encourage the adoption of circular economy practices, such as recycling and reusing materials.

Environmental and Public Health Benefits

The reduction of air pollutants, such as NOx, SO_2 and particulate matter, has led to significant improvements in air quality and public health. Fewer pollutants in the air reduce the incidence of respiratory and cardiovascular diseases. Improving air quality also has economic benefits, such as reduced healthcare costs and increased labor productivity due to better overall health of the population.

Conclusion

Emissions standards are essential for the transition to a sustainable future, promoting clean technologies, improving air quality and protecting public health. Through coordinated and ambitious efforts, governments and industries can achieve significant reductions in emissions, moving towards a cleaner and healthier world for all. With the implementation of these regulations, technological innovation and the adoption of more sustainable practices can be encouraged in various sectors, thus ensuring a long-term positive impact on the environment and society.

Future perspectives: When will we stop using combustion cars?

The end of the era of internal combustion cars seems to be getting closer, driven by government policies, technological advances, and changes in consumer preferences. However, determining the exact moment when we will stop using these vehicles is complex and depends on multiple factors. Current trends, global policies, and future prospects are shaping the fate of combustion cars significantly.

At the global level, government policies play a crucial role in this transition. In Europe, the European Union has set ambitious targets to reduce CO_2 emissions from cars. Several European countries, including Norway, the United Kingdom, and France, have announced plans to ban the sale of new internal combustion cars between 2025 and 2040. These bans are driving automakers to accelerate the transition to electric and hybrid vehicles, developing new technologies and expanding their offering of sustainable models. In Norway, for example, nearly 80% of new cars sold in 2021 were electric, a testament to the impact of government policies and economic incentives.

In North America, some U.S. states, such as California, have announced plans to ban the sale of new internal combustion cars by 2035. At the federal level, the Biden administration has proposed measures to increase EV adoption and improve charging infrastructure. In 2021, the administration launched a plan to install 500,000 charging stations across the country by 2030. Canada has also announced its intention to ban the sale of new internal combustion cars by 2035, supporting this transition with favorable incentives and policies.

In Asia, China, the world's largest automotive market, has set targets for new energy vehicles (NEVs) to account for a significant share of car sales by 2035. Chinese cities are already implementing restrictions to limit the use of combustion vehicles. Japan has also set a target to phase out internal combustion cars in favor of electric and hybrid vehicles by the mid-2030s. The city of Shenzhen, for example, has fully electrified its bus fleet, demonstrating China's commitment and ability to lead this transition.

Technological advances are also accelerating this transition. Improvements in the energy density of lithium-ion batteries, along with reduced costs and faster charging times, are making electric vehicles more competitive with internal combustion vehicles. Solid-state batteries promise to offer even higher energy densities, faster charging times, and improved safety, which could further accelerate the adoption of electric vehicles. Companies such as Toyota and QuantumScape are leading the development of these technologies, with plans to commercialize them in the coming years.

The expansion of charging infrastructure is another crucial factor. Many governments are investing in the expansion of charging stations, setting targets for installing charging stations on roads, urban and rural areas. Partnerships between governments and private companies are accelerating the deployment of fast and ultra-fast charging stations. Innovations in charging technologies, such as wireless charging and ultra-fast charging, improve the convenience and viability of these vehicles for long journeys. In 2021, Tesla announced the expansion of its Supercharger network, with more than 30,000 stations worldwide, allowing a vehicle to be charged in less than 30 minutes.

Changes in consumer preferences are also playing a role. Growing awareness of climate change and its impacts has led to increased interest in sustainable transport solutions. Information and education campaigns on the benefits of electric vehicles are helping to change perceptions and increase the adoption of these vehicles. Consumers are increasingly inclined to choose electric vehicles due to their environmental benefits and government incentives. In a recent survey, 55% of consumers in the United States expressed interest in considering an electric vehicle for their next purchase.

Tax subsidies and rebates are a key tool for reducing the upfront cost of electric vehicles, making them more affordable for consumers. Long-term benefits, such as lower maintenance costs and lower fuel costs, also appeal to consumers looking for economical and efficient options. A study by Consumer Reports found that EV owners save an average of $4,600 in maintenance and fuel costs over the life of the vehicle compared to internal combustion vehicles.

However, there are challenges and barriers that still need to be overcome. The availability of charging stations varies significantly between urban and rural regions, which can be a barrier to widespread adoption. A lack of standardization in connectors and charging protocols can hinder interoperability and convenience for users. The increasing demand for energy due to EV charging can put a strain on power grids, especially in areas with aging electrical infrastructure.

Concerns about the limited range of electric vehicles remain a barrier for some consumers, although technological improvements are mitigating this problem. Changing the mindset of consumers and businesses requires continued efforts in education, incentives, and demonstration of long-term benefits.

An inspiring story in this context is that of Elon Musk and Tesla. Musk, driven by his vision for a sustainable future, has brought Tesla to the forefront of the electric vehicle revolution. Despite countless initial challenges and skepticisms, Tesla has managed to not only prove that electric vehicles can be practical and desirable, but also to set new standards for performance and range. With models like the Tesla Model

S and Model 3, the company has shown that electric vehicles can outperform internal combustion vehicles in many ways.

The transition to a future without internal combustion cars is underway, driven by government policies, technological advancements, and changes in consumer preferences. While the road is not without its challenges, globally coordinated efforts are hastening the end of the era of internal combustion vehicles. While it is difficult to predict exactly when we will stop using these cars altogether, current trends suggest that the 2030s will be a watershed period in this transition, marking a significant shift towards cleaner and more sustainable mobility.

Chapter 7: Looking Forward: The Energy Revolution

Renewable energies and their role in the energy transition

The energy transition to a more sustainable future is defined as the shift from the use of fossil fuels to renewable energy sources, essential for reducing greenhouse gas emissions. This concept, which has gained momentum in recent decades, has its roots in the Industrial Revolution when coal replaced wood. During the 20th century, oil and natural gas became the main sources of energy, creating a global economy dependent on these fossil fuels. Over time, growing concern about climate change has catalyzed a new phase of energy transition, focusing on the adoption of renewable energies such as solar, wind, hydroelectric and geothermal.

Solar energy is one of the cornerstones of this transition. Advances in photovoltaic panels, which convert sunlight into electricity through the photovoltaic effect, have increased their efficiency and accessibility. These panels are becoming more common in homes, businesses, and large-scale solar power plants. Solar thermal power plants, which use mirrors to concentrate sunlight on a receiver, have improved in performance and economic viability. These plants convert solar energy into heat, which is then used to generate electricity. Solar energy offers the advantage of being an inexhaustible source and produces no emissions during its operation. However, it faces challenges such as intermittency and the need for adequate space for its installation.

Wind energy also plays a crucial role in the energy transition. Wind turbines convert the kinetic energy of the wind into electricity by rotating their blades. Wind farms, both onshore and offshore, take advantage of stronger and more constant wind currents. Wind energy produces no emissions during its operation and has low operating costs once the wind turbines are installed. However, their production can be unpredictable and can have environmental and social impacts, such as wildlife and noise. Studies from Stanford University have shown that the expansion of wind power could significantly reduce global CO_2 emissions, if deployed on a large scale.

Hydropower, which uses moving water to generate electricity, has been a stable energy source for decades. Large dams store water and release it in a controlled manner to drive turbines, generating electricity. In addition to providing a constant and controllable source of energy, dams can act as natural energy storage. However, the construction of large dams can have significant impacts on ecosystems and local communities. Micro-hydroelectric plants, which take advantage of smaller rivers and streams, are a less invasive and more suitable alternative for certain regions. According to MIT studies, micro-hydroelectric plants could play a crucial role in generating sustainable energy in rural areas.

Geothermal energy, which harnesses the Earth's internal heat, provides a constant and predictable source of energy. Geothermal plants use this heat to generate electricity or provide heating. Geothermal energy is particularly viable in regions with high geothermal activity. Geothermal heat pumps, which use the constant temperature of the subsurface for heating and cooling, are gaining popularity in residential and commercial applications. Although its initial installation can be expensive, geothermal energy offers low operating costs and a reliable long-term source of energy.

Renewable energies not only help reduce greenhouse gas emissions, but also diversify the energy matrix, reducing dependence on imported fossil fuels and increasing energy security. These energies also drive economic development and create jobs in the construction, operation and maintenance of facilities. According to the International Renewable Energy Agency (IRENA), the renewable energy sector could generate more than 42 million jobs globally by 2050.

However, the implementation of renewable energy faces significant challenges. The integration of these sources into the electricity grid requires proper management of the intermittency and variability of production. Energy storage systems, such as large-scale batteries, are essential to manage this variability. Smart grids allow for better management of electricity demand and supply, optimizing power distribution and improving the efficiency of the electricity system.

Financing and political support are crucial to boost investment in renewable infrastructure. Subsidies, tax credits, and other government incentives can encourage investment in renewable energy and support its deployment. In addition, the creation of green investment funds and the participation of the private sector are essential to mobilise the financial resources needed for the energy transition.

Education and community participation are also essential. Awareness campaigns on the benefits of renewable energy and the need for an energy transition can increase public support. Involving local communities in the development of renewable energy projects can ensure their long-term success and sustainability. Education and training in renewable energy technologies develop the skilled workforce needed to build and maintain these facilities.

Looking ahead, technological innovations will continue to drive the energy transition. Advances in energy storage, such as next-generation batteries and green hydrogen, will improve the storage capacity and efficiency of renewables. Green hydrogen, produced by electrolysis with renewable energy, is a promising solution for large-scale energy storage and use in hard-to-electrify sectors. Ocean energy, such as tidal energy and wave energy, also has the potential to provide a significant source of renewable energy in the future.

At the global level, agreements such as the Paris Agreement and the UN Sustainable Development Goals promote the adoption of policies and technologies that favour the energy transition. National energy plans and net-zero commitments reflect the global ambition to reduce emissions and adopt clean energy. According to studies by the University of California, the mass adoption of renewable energy could reduce global CO_2 emissions by 70% by 2050.

Before we consider migrating to other planets, we need to focus on maintaining and improving the planet we live on. Renewable energy offers the opportunity to build a more sustainable future on Earth, ensuring a clean and healthy planet for future generations. The energy transition not only represents a challenge, but also an opportunity to innovate, grow and create a better world for all. With a coordinated global effort, we can achieve a cleaner, safer, and more sustainable future, leveraging renewable energy to meet the energy and climate challenges of our era.

Alternatives to the electric car: Hydrogen and beyond

As the electric car consolidates itself as a sustainable and viable option, other promising alternatives are emerging that could complement and diversify the clean transport landscape. Hydrogen, for example, is emerging as a key energy source for the vehicles of the future. Hydrogen vehicles use fuel cells to convert hydrogen into electricity. In this process, hydrogen combines with oxygen inside the fuel cell, generating electricity, water, and heat as byproducts. This method is highly efficient and produces no polluting emissions, emitting only water vapor.

The production of hydrogen can be done by electrolysis, a process that breaks down water into hydrogen and oxygen using electricity. If this electricity comes from renewable sources, the resulting hydrogen is completely carbon-free. Another technique is natural gas reforming, which is currently the most common, although it produces CO_2 emissions. However, carbon capture and storage (CCS) can mitigate these emissions.

Hydrogen vehicles have several significant advantages. They offer a longer range than battery electric vehicles, similar to that of internal combustion vehicles. In addition, the hydrogen refueling time is very short, usually less than 10 minutes, comparable to filling up a gas tank. Hydrogen vehicles hold particular promise for heavy-duty transport such as trucks, buses and trains, where electric batteries may be less practical due to their weight and recharging time.

However, the mass adoption of hydrogen vehicles faces several challenges. One of the main barriers is the lack of replenishment infrastructure. Significant investment is required to build a network of hydrogen stations. In addition, the cost of producing and distributing hydrogen is currently higher than that of fossil fuels and grid electricity, although costs are expected to decrease with time and economies of scale. The hydrogen conversion chain (production, storage, transportation, and conversion into electricity) can also be less efficient than electric batteries due to energy losses at each stage of the process.

Apart from hydrogen, other emerging technologies are revolutionizing sustainable mobility. Advanced biofuels, for example, are produced from renewable feedstocks such as agricultural waste, algae and used oils. These biofuels can significantly reduce CO_2 emissions compared to fossil fuels and can be used in existing internal combustion engines with little or no modifications. They are especially attractive for air and sea freight, where electric alternatives are less viable due to power and range needs.

One company that is making significant strides in this field is Repsol. They have developed a renewable fuel that is produced from biomass and waste, which not only reduces CO_2 emissions, but is also compatible with current internal combustion engines. This innovation is an example of how biofuels can be integrated into today's market without requiring major changes to existing infrastructure.

Solar vehicles represent another exciting innovation. They integrate photovoltaic panels into their structure, allowing electricity to be generated while they are in motion or parked. Although solar panels cannot cover all of a vehicle's energy needs, they can extend its range and reduce the frequency of charging. The efficiency of solar panels and the surface area available in a vehicle limit the amount of energy that can be generated, so continued research into more efficient and lighter solar panels is crucial to improving the viability of these vehicles.

Infrastructure development is critical to the adoption of these emerging technologies. Governments can play a key role in providing incentives, subsidies and supportive policies. Public-private partnerships can accelerate infrastructure development, sharing costs and risks. For hydrogen vehicles, it is essential to develop a network of refueling stations, including stations in urban areas, transport corridors and rural areas. For solar vehicles, charging infrastructure can include parking lots with solar roofs and charging stations equipped with solar panels.

Public education and awareness are also crucial. Educational campaigns can increase public awareness of the benefits and possibilities of emerging technologies, encouraging their acceptance and adoption. Pilot projects and public demonstrations can show the workings and benefits of new technologies, helping to overcome skepticism and resistance to change. Education and training in new technologies are essential to develop the workforce needed to build, maintain, and operate advanced infrastructure and vehicles. Fostering the transfer of knowledge and best practices between countries and communities can accelerate the adoption of emerging technologies.

Alternatives to electric cars, such as hydrogen vehicles and other emerging technologies, play a crucial role in the transition to sustainable mobility. Each technology has its own advantages, challenges, and areas of application, and a combination of these solutions is likely to be necessary to address the diverse needs of global transportation. Infrastructure development, government support, investment in research, and public education are key factors for the success of these technologies. Looking ahead, it is essential to continue to explore and support these alternatives to achieve a cleaner, safer and more sustainable energy future. The energy revolution is not limited to a single solution, but encompasses a diverse and dynamic ecosystem of innovations that, together, can transform transport and contribute significantly to the fight against climate change.

In addition, renowned universities have been researching and supporting these technologies. For example, the Massachusetts Institute of Technology (MIT) has conducted studies demonstrating how the adoption of renewable energy and clean technologies can boost economic growth and job creation. Advances in hydrogen and biofuels not only offer clean energy solutions, but can also revitalize local and national economies by creating new industries and markets.

In conclusion, the transition to sustainable mobility is a complex challenge that requires a multifaceted approach. Emerging technologies such as hydrogen vehicles, advanced biofuels, and solar vehicles offer promising solutions that can complement and enhance the impact of electric cars. With the right support, these technologies can play a crucial role in building a cleaner and more sustainable energy future. By leveraging these innovations, we can not only meet the current challenges of climate change, but also ensure a healthier and more livable planet for future generations.

The future of sustainable mobility

Sustainable mobility is presented as a fundamental pillar in the fight against climate change and environmental pollution. As we move toward a cleaner, more efficient future, a variety of technologies and innovations are transforming the way we get around. This evolution is not only based on the electrification of vehicles, but also on the implementation of smart solutions and sustainable urban planning that facilitate a more efficient and environmentally friendly transport system.

In cities, electrified public transport has proven to be an effective solution for reducing greenhouse gas emissions. Electric buses, for example, offer significant advantages: they are quieter, require less maintenance and, most importantly, do not emit local pollutants. Cities such as Shenzhen in China have led this transformation, fully electrifying their bus fleet and demonstrating the environmental and economic benefits of this transition. This success has inspired other cities to follow suit, increasing their fleets of electric buses.

Electric trains and trams also play a crucial role in urban transport. With their high energy efficiency and the possibility of being powered by renewable energy sources, these mass transit systems are ideal for dense urban areas. Many cities already have existing infrastructure for electric trains and trams, making it easier to expand and modernize. These networks not only improve urban mobility, but also contribute significantly to the reduction of congestion and pollution.

The rise of bike and scooter sharing systems has added an extra layer of flexibility and accessibility to urban transportation. Bicycles, both electric and traditional, offer an affordable and convenient transportation option, reducing car use and associated emissions. In addition, they encourage physical activity, contributing to public health. Electric scooters, on the other hand, are an ideal micromobility solution for short journeys in urban environments. Although their rapid proliferation has posed challenges in terms of regulation and safety, their potential to reduce congestion and emissions is undeniable.

Innovation in battery technology has been a key factor in the advancement of electric vehicles. Solid-state batteries promise higher energy density and greater safety, which could significantly increase the range of electric vehicles and reduce charging times. In addition, improving battery recycling techniques is crucial to minimising the environmental impact of electric vehicles. Companies and researchers are developing efficient methods to recover valuable materials from used batteries, which not only reduces waste, but can also reduce the costs of producing new batteries.

Wireless and ultra-fast charging are other areas of innovation that are transforming electric mobility. Wireless charging eliminates the need for plugs and cables, allowing for more convenient and efficient charging in urban and residential environments. Ultra-fast charging stations, on the other hand, make it possible to recharge an electric vehicle in minutes instead of hours, which is essential for the mass adoption of these vehicles. However, the implementation of ultra-fast charging stations requires significant improvements in electrical infrastructure to handle the high energy demand.

Autonomous and connected vehicles represent the next big revolution in sustainable mobility. Equipped with advanced sensors, artificial intelligence and machine learning algorithms, these vehicles can navigate safely and efficiently, significantly reducing traffic accidents caused by human error. In addition, autonomous vehicles can offer on-demand transportation services, optimizing routes and reducing traffic congestion. Vehicle-to-vehicle (V2V) and vehicle-to-infrastructure (V2I) communication can further improve the efficiency and safety of urban transport.

Sustainable urban planning is essential to support these technological innovations. The design of compact and dense cities reduces the need for car travel, encouraging the use of public transport, bicycles and pedestrian mobility. The creation of low-emission zones in cities can incentivise the use of electric vehicles and improve urban air quality. In addition, the integration of multimodal transport networks, which combine different forms of sustainable transport, facilitates access and mobility for all citizens. The expansion of green public spaces and infrastructure for pedestrians and cyclists also promote a healthy and accessible urban environment.

Incentive and regulatory policies play a crucial role in promoting sustainable mobility. Subsidies and tax incentives for the purchase of electric vehicles and other sustainable means of transport can accelerate the transition to cleaner mobility. Incentives for the installation of charging stations and other related services are essential for the development of adequate infrastructure. In addition, implementing stricter emissions standards and energy efficiency regulations can significantly reduce air pollution and contribute to the reduction of greenhouse gas emissions.

A notable example of innovation in sustainable fuels is the development of advanced biofuels by companies such as Repsol. These fuels are produced from renewable raw materials, such as agricultural waste and used oils, rather than food crops. Biofuels can significantly reduce CO_2 emissions and are compatible with existing internal combustion engines, facilitating their adoption and transition. In addition, they are particularly attractive for air and sea freight, where electric alternatives are less viable.

Looking to the future, the combination of emerging technologies and comprehensive planning and regulation strategies can transform urban mobility. Education and public awareness are essential to increase the acceptance and adoption of new technologies. Educational campaigns can increase public awareness of the benefits of emerging technologies, while pilot projects and public demonstrations can show their operation and benefits. Education and training in new technologies are also crucial to developing the workforce needed to build, maintain, and operate advanced infrastructure and vehicles.

Sustainable mobility is not only a matter of technology, but also of policies and planning. A comprehensive approach that combines diverse technology solutions and policy strategies can create an efficient, accessible, and environmentally friendly transportation system. From the electrification of public transport to the development of autonomous and connected vehicles, the opportunities to transform mobility are vast. With coordinated efforts and a long-term vision, we can build a transportation system that not only improves our lives, but also protects the planet for future generations.

The transition to sustainable mobility is a complex challenge that requires a multifaceted approach. Emerging technologies such as hydrogen vehicles, advanced biofuels, and solar vehicles offer promising solutions that can complement and enhance the impact of electric cars. With the right support, these technologies can play a crucial role in building a cleaner and more sustainable energy future.

By leveraging these innovations, we can not only meet the current challenges of climate change, but also ensure a healthier and more livable planet for future generations. The energy revolution is not limited to a single solution, but encompasses a diverse and dynamic ecosystem of innovations that, together, can transform transport and contribute significantly to the fight against climate change.

Conclusions: Towards a Greener and More Peaceful World

Throughout this book, we have explored the various facets of the energy revolution, from historical dependence on oil to the emergence of clean and sustainable technologies. Now, it is crucial to reflect on how these changes can lead us towards a greener and more peaceful world. In these conclusions, we will look at the environmental, economic, and social benefits of this transition, and how each of us can contribute to a more sustainable future.

Reducing emissions and improving air quality are perhaps the most visible and immediate benefits of the transition to sustainable transport. The mass adoption of electric and hydrogen-powered vehicles has the potential to significantly reduce emissions of greenhouse gases and other pollutants. Imagine a city where most vehicles are quiet and don't emit toxic fumes, a place where respiratory and cardiovascular diseases decrease dramatically, and the air quality is such that the sky is bluer and wildlife flourishes again.

Public transport also plays a crucial role in this vision. The electrification of buses, trains, and trams not only reduces emissions, but also improves energy efficiency and reduces urban noise. Shenzhen, in China, has already managed to fully electrify its bus fleet, proving that this dream can come true. Imagine traveling on a clean and quiet bus, knowing that your trip is not contributing to global warming or the pollution of your city.

But the benefits of renewable energy go beyond transportation. Energy sources such as solar, wind, and geothermal produce electricity without emitting greenhouse gases, which is essential for mitigating climate change. These energies also help preserve biodiversity and ecosystems, by reducing the exploitation of natural resources and pollution. Solar energy, for example, is transforming entire landscapes, covering rooftops and fields with panels that convert sunlight into clean electricity.

In the economic sphere, the transition to a green economy is creating new job opportunities and promoting sustainability. The expansion of renewable energies and sustainable mobility is generating millions of jobs in sectors such as the manufacture, installation, maintenance and management of green infrastructure. Battery recycling and the circular economy are also driving job creation, promoting economic sustainability and reducing waste. Companies like Repsol are leading the way in the production of renewable fuels, demonstrating that it is possible to combine innovation and environmental responsibility.

In addition, the operating and maintenance costs of electric and hydrogen-powered vehicles are significantly lower compared to internal combustion vehicles. This translates into considerable savings for consumers and businesses. Energy price stability is also a key benefit, as diversification of energy sources and independence from fossil fuels protect economies from oil and gas market volatility.

From a social perspective, the transition to clean energy and sustainable mobility improves quality of life and promotes equity.

Reducing air and water pollution improves public health, decreasing the incidence of pollution-related diseases and increasing life expectancy. Imagine living in a city where the air is clean and breathable, where children can play outside without fear of the harmful effects of pollution.

It is crucial that the energy transition is inclusive and equitable, ensuring that economic and environmental benefits reach all communities, especially those that have been most affected by pollution and resource exploitation. Involving communities in decision-making on renewable energy and sustainable mobility projects ensures that their needs and concerns are addressed, promoting environmental justice.

Despite the many benefits, the transition to a sustainable future is not without its challenges. Developing robust infrastructure for renewable energy and sustainable vehicles requires significant investments. Governments, the private sector and financial institutions must work together to mobilize the necessary resources. The implementation of smart grids and advanced energy storage solutions is essential to manage the variability of renewable sources and ensure a constant and reliable supply of electricity.

Technological innovation must continue to improve efficiency, reduce costs, and increase the accessibility of clean technologies. This includes the development of new energy storage technologies, advanced materials, and sustainable production methods. In addition to mitigating climate change, it is crucial to develop adaptation strategies to cope with unavoidable impacts, such as extreme weather events and changes in weather patterns.

International collaboration is key to achieving sustainability goals. Commitments under the Paris Agreement and the UN Sustainable Development Goals provide a comprehensive framework for addressing environmental, economic and social challenges, promoting sustainable and inclusive development globally. Cities and regions can play a crucial role in implementing sustainable policies and projects, tailored to their specific contexts. Successful local initiatives can serve as role models and be replicated elsewhere.

Fostering public awareness and education about sustainability and climate change is essential to mobilize community action and support for green policies. Educational campaigns and awareness programs can increase public awareness of the benefits of emerging technologies and the need for an energy transition. Education and training in new technologies are also crucial to developing the workforce needed to build, maintain, and operate advanced infrastructure and vehicles.

Ultimately, the path to a greener and more peaceful world is full of challenges, but also immense opportunities. The transition to clean energy and sustainable mobility is not only an environmental necessity, but also an opportunity to improve the quality of life, promote economic development and ensure social justice. Each of us has a role to play in this transformation. From supporting sustainable policies and practices to adopting clean technologies in our daily lives, our collective actions can make a difference. With determination, innovation and global collaboration, we can build a future where clean energy and sustainable mobility are not only possible, but the norm, creating a healthier, fairer and more prosperous planet for all.

The energy revolution is not limited to a single solution, but encompasses a diverse and dynamic ecosystem of innovations that, together, can transform transport and contribute significantly to the fight against climate change. By leveraging these innovations, we can not only meet the current challenges of climate change, but also ensure a healthier and more livable planet for future generations. The transition to clean energy and sustainable mobility is an exciting and challenging journey, full of possibilities for a better future.

Key lessons and conclusions

Throughout this book, we have gone on a fascinating journey towards a cleaner and more sustainable future, discovering how the transition to renewable energy and sustainable vehicles can transform our planet. The adoption of clean energies, such as solar, wind and geothermal, together with electric mobility and the use of hydrogen, are key to reducing greenhouse gas emissions. These technologies not only improve air quality, but also mitigate climate change and preserve biodiversity.

The economic impact of this transition is equally significant. The green economy is generating a wide range of new job opportunities in emerging sectors such as clean technology manufacturing, renewable energy infrastructure installation and recycling. Innovative companies are leading the way, developing more efficient and sustainable methods, such as Repsol's renewable fuel. These developments not only promote economic sustainability, but also ensure the stability of energy prices, protecting economies from the volatility of fossil fuel markets.

From a social point of view, the benefits of the energy transition are profound. The improvement in public health is one of the most evident, with a significant reduction in respiratory and cardiovascular diseases thanks to the decrease in air pollution. In addition, access to clean and affordable energy improves the quality of life, especially in rural and disadvantaged communities. It is crucial that this transition is inclusive and equitable, ensuring that all communities, especially those most affected by pollution and resource exploitation, benefit from technological and economic advances.

Innovation and technological development are fundamental drivers of this transformation. Advances in batteries, such as solid-state batteries, promise higher energy density, safety and faster charging times, while hydrogen is emerging as a promising solution for heavy-duty transport. Ultra-fast wireless charging technology is improving the convenience and efficiency of electric vehicles, making sustainable mobility more accessible to everyone.

The development of robust infrastructure and public policy support are essential to facilitate this transition. Investment in smart grids and energy storage solutions, as well as charging infrastructure for electric vehicles, is crucial to ensure a constant and reliable supply of electricity. Subsidies and tax incentives for the purchase of sustainable vehicles and the installation of charging infrastructure also play a key role in accelerating the adoption of these technologies.

Global collaboration is another fundamental pillar on the path to a greener and more peaceful future. International agreements, such as the Paris Agreement, and the UN Sustainable Development Goals, provide a comprehensive framework for addressing environmental, economic, and social challenges. Cooperation between countries is essential to share knowledge, technologies and resources, enabling a faster and more efficient energy transition. Cities and regions can also play a crucial role in implementing sustainable policies and projects, tailored to their specific contexts. Successful local initiatives can serve as role models and be replicated elsewhere.

Fostering public awareness and education about sustainability and climate change is essential to mobilize community action and support for green policies. Educational campaigns and awareness programs can increase public awareness of the benefits of emerging technologies and the need for an energy transition. Education and training in new technologies are also crucial to developing the workforce needed to build, maintain, and operate advanced infrastructure and vehicles.

Ultimately, the path to a greener and more peaceful world is full of challenges, but also immense opportunities. The transition to clean energy and sustainable mobility is not only an environmental necessity, but also an opportunity to improve the quality of life, promote economic development and ensure social justice. Each of us has a role to play in this transformation. From supporting sustainable policies and practices to adopting clean technologies in our daily lives, our collective actions can make a difference. With determination, innovation and global collaboration, we can build a future where clean energy and sustainable mobility are not only possible, but the norm, creating a healthier, fairer and more prosperous planet for all.

The energy revolution is not limited to a single solution, but encompasses a diverse and dynamic ecosystem of innovations that, together, can transform transport and contribute significantly to the fight against climate change. By leveraging these innovations, we can not only meet the current challenges of climate change, but also ensure a healthier and more livable planet for future generations. The transition to clean energy and sustainable mobility is an exciting and challenging journey, full of possibilities for a better future.

Implications for the future

The transition to a more sustainable world has profound implications for our future. In this chapter, we explore how energy transformation and sustainable mobility can radically change our lives. As the world faces the challenge of climate change, the need to adopt clean energy and advanced technologies becomes more urgent. From the implementation of renewable energy to the electrification of transportation, these changes are aimed at reducing greenhouse gas emissions and improving air quality. Solar, wind, hydroelectric and geothermal energy are positioning themselves as the fundamental pillars of this change, allowing a significant decrease in our dependence on fossil fuels.

Technological innovation plays a crucial role in this transformation. Advances in energy storage, such as lithium-ion batteries and hydrogen systems, are essential to effectively integrate renewables into our electricity grids. In addition, smart grids will facilitate a more efficient distribution of energy, managing demand and supply in real time.

In the field of sustainable mobility, electric and hydrogen vehicles are revolutionizing transportation. The mass adoption of these vehicles will not only reduce air pollution, but will also decrease CO_2 emissions from the transport sector. Electric vehicles offer numerous advantages, such as lower operating and maintenance costs, as well as quiet and efficient performance. However, for widespread adoption, it is crucial to develop a suitable charging infrastructure. Fast and ultra-fast charging stations are necessary to reduce charging times and make electric vehicles more convenient for users.

On the other hand, hydrogen vehicles present a promising solution, especially for heavy-duty and long-distance transport. They use fuel cells to convert hydrogen into electricity, with the only byproduct being water vapor. This process is highly efficient and does not produce polluting emissions. However, hydrogen refueling infrastructure is still limited and requires significant investments for expansion.

In addition to electric and hydrogen vehicles, micromobility solutions such as shared bikes and electric scooters are gaining popularity in cities around the world. Not only are these modes of transportation clean and efficient, but they also help reduce traffic congestion and promote a more active lifestyle. The electrification of public transport is also a key element in sustainable mobility. Cities such as Shenzhen in China have demonstrated the benefits of fully electrifying their bus fleets, significantly reducing emissions and improving urban air quality.

Energy transformation and sustainable mobility not only have environmental benefits, but also economic and social ones. The green economy is generating new employment opportunities in sectors such as renewable energy, sustainable construction and electric mobility. The transition to clean technologies not only promotes economic growth, but also improves the competitiveness of economies. Green jobs are not only good for the environment, but they also offer opportunities for professional development and improve quality of life.

Improving public health is another important benefit of the energy transition. Reducing air and water pollution contributes to better public health, reducing the incidence of respiratory and cardiovascular diseases. In addition, ensuring that the benefits of the energy transition reach all communities is crucial to promoting equity and environmental justice. Community participation in decision-making on renewable energy and sustainable mobility projects ensures that their needs and concerns are addressed, promoting a fair distribution of benefits.

International cooperation is essential to achieve a successful energy transition. Commitments under agreements such as the Paris Agreement are critical to coordinating efforts and achieving global climate goals. The transfer of clean technologies and best practices between countries accelerate the energy transition globally. In addition, providing financing and technical assistance to developing countries is vital for them to implement sustainable energy solutions and adapt to climate change. Promoting sustainable development in all regions of the world is critical to ensuring a prosperous and equitable future for all.

The future of sustainable energy and mobility is full of promise and opportunity. With global engagement, strategic investments, and a clear vision, we can build a greener, more peaceful world. The transition to a sustainable future requires the collaboration of everyone: governments, businesses, communities and citizens. Together, we can meet the challenges of climate change, improve the quality of life, and ensure a healthy planet for future generations. The energy revolution is not only an environmental necessity, but also an opportunity to improve our lives and create a more just and equitable future.

Each of us has a role to play in this transformation. From supporting sustainable policies and practices to adopting clean technologies in our daily lives, our collective actions can make a difference. With determination, innovation and global collaboration, we can build a future where clean energy and sustainable mobility are not only possible, but the norm, creating a healthier, fairer and more prosperous planet for all.

Policy Recommendations and Future Research

As we enter the era of sustainability, effective policy implementation and continued research into emerging technologies become critical pillars for securing a greener and more peaceful future. Imagine a world where cities are free of smog, where energy flows cleanly and efficiently, and where transportation is quiet and non-polluting. This is not a distant dream, but an achievable goal with the right strategies.

The first step towards this vision is to transform our energy and mobility policies. Financial incentives, such as subsidies and tax breaks, play a crucial role in the adoption of clean technologies. Let's imagine a scenario where every household has access to affordable solar panels and subsidized electric vehicles, significantly reducing greenhouse gas emissions. In addition, financing green infrastructure projects, such as electric vehicle charging stations and renewable energy plants, is essential for their development.

Regulations and standards are also critical. Implementing strict emission standards for vehicles and industrial sectors will not only help reduce pollution, but will also drive technological innovation. Setting energy efficiency standards for buildings, appliances, and industrial machinery is another crucial step. Imagine living in a city where every building is designed to maximize energy efficiency, reducing consumption and emissions.

Infrastructure development is another critical area. Expanding the network of charging stations for electric and hydrogen vehicles, ensuring their accessibility in urban and rural areas, is vital. Imagine a network of charging stations on every corner of the city, making EV charging as easy as filling up a gas tank. Investing in the electrification of public transport is also crucial to reduce reliance on private transport and emissions. Think electric buses and trains that offer clean and efficient transportation.

The integration of renewable energies into our electricity grid is essential. Developing smart grids that optimize energy distribution and manage demand efficiently can transform the way we consume energy. In addition, promoting research and implementation of energy storage technologies, such as advanced batteries and hydrogen systems, will ensure a steady supply of renewable energy. Imagine a world where solar energy collected during the day is efficiently stored for use at night, ensuring an uninterrupted supply of clean energy.

Research into emerging technologies is also vital. Solid-state batteries, for example, could revolutionize electric mobility by offering higher energy density, greater safety, and a longer lifespan. Hydrogen systems also offer a promising clean and efficient energy vector.

Research in ocean energy, such as tidal and wave energy, has great untapped potential. Artificial photosynthesis is another fascinating area, which seeks to create clean fuels from sunlight, water and CO_2, emulating the natural process of plants.

The social and economic impacts of the energy transition are also significant. The green economy is creating new jobs in sectors such as renewable energy, sustainable construction and electric mobility. In addition, promoting equity in the energy transition is crucial to ensure that everyone benefits from technological advances. Imagine entire communities thriving thanks to the creation of green jobs and access to clean, affordable energy.

International cooperation is essential to achieve these goals. Commitments under agreements such as the Paris Agreement are essential to coordinate global efforts and achieve climate goals. The transfer of clean technologies and best practices between countries will accelerate the energy transition. Providing financing and technical assistance to developing countries is vital for them to implement sustainable energy solutions and adapt to climate change.

In conclusion, the future of sustainable energy and mobility is full of promise and opportunities. With global engagement, strategic investments, and a clear vision, we can build a greener, more peaceful world. The transition to a sustainable future requires the collaboration of everyone: governments, businesses, communities and citizens. Together, we can meet the challenges of climate change, improve the quality of life, and ensure a healthy planet for future generations. The energy revolution is not only an environmental necessity, but also an opportunity to improve our lives and create a more just and equitable future. Each of us has a role to play in this transformation. From supporting sustainable policies and practices to adopting clean technologies in our daily lives, our collective actions can make a difference. With determination, innovation and global collaboration, we can build a future where clean energy and sustainable mobility are not only possible, but the norm, creating a healthier, fairer and more prosperous planet for all.

References: Sources and Recommended Reading

Abello Avila, A., González Melo, H., & Oviedo Poveda, I. (2021). The pedagogical approach to the memory of forced displacement with children: a commitment to the recognition of otherness in the classroom. *Revista Educación, 45*(2), 449-467. Retrieved from https://www.researchgate.net/publication/353265869_El_abordaje_pedagogico_de_la_memoria_del_desplazamiento_forzado_con_ninos_y_ninas_una_apuesta_para_el_reconocimiento_de_la_otredad_en_el_aula

Abrahamian, E. (2021). *Oil Crisis in Iran: From Nationalism to Coup D'etat.* Cambridge University Press.

UN Refugee Agency. (2022). *Displacements in the world, causes and consequences.* Retrieved from https://eacnur.org/es/desplazamientos-en-el-mundo

International Energy Agency. (2007). *World Energy Outlook 2007.* Retrieved from https://www.iea.org/reports/world-energy-outlook-2007

Agüero, C. (2022). *The destruction of Iraq's historical and cultural heritage during 2003.* Retrieved from https://www.aacademica.org/2.congreso.internacional.de.ciencias.humanas/387

Aguilar Valdera, E. (2017). Analysis of Kia Picanto 1.0mpi engine tests to homologate Fifth-generation LPG conversion kit. Retrieved from https://repositorio.ucv.edu.pe/handle/20.500.12692/25117

IEA. (2022). Exploring the impacts of the Covid-19 pandemic on global energy markets, energy resilience, and climate change. Retrieved from https://www.iea.org/topics/covid-19

Altamirano, T. (2021). *Altamirano, T. (2021). Environmental refugees: climate change and forced migration.* PUCP Publishing Fund.

Álvarez, C. (2014, January 31). *Which pollutes more, an electric car or a petrol car?* Retrieved from El País weekly Blogs: https://blogs.elpais.com/eco-lab/2014/01/que-contamina-mas-un-coche-electrico-o-uno-de-gasolina.html

Alvarez, R. (2013). Structure and recomposition of the world automotive industry. Opportunities and perspectives for Mexico. *Economía UNAM, 10*(30), 75-92. Retrieved from https://www.elsevier.es/es-revista-economia-unam-115-articulo-estructura-recomposicion-industria-automotriz-mundial--S1665952X13722047

Amadoz, S. (2023, January 10). *How much CO2 does an electric car really emit? This tool calculates it.* Retrieved from El País Motor: https://motor.elpais.com/actualidad/cuanto-co2-emite-de-verdad-un-coche-electrico-esta-herramienta-lo-calcula/

Ameer H, A.-R. (2023). The Impacts of Petroleum on Environment. *IOP Conference Series: Earth and Environmental Science.* Retrieved from https://iopscience.iop.org/article/10.1088/1755-1315/1158/3/032014/meta

Aminudin, M., Kamarudin, S., Lim, B., Majilan, E., Masdar, M., & Shaari, N. (2023). An overview: Current progress on hydrogen fuel cell vehicles. *International Journal of Hydrogen Energy, 48*(11), 4371-4388. Retrieved from https://doi.org/10.1016/j.ijhydene.2022.10.156

Aranda, D. (2015). *Scorched earth: Oil, soybeans, pulp mills and mega-mining. Radiography of Argentina in the 21st century.* South american.

Arcos Vargas, Á. (2022). The electric car: Strengths and weaknesses for its expansion. *Papeles de Economía Española* (171), 63-175. Retrieved from https://www.proquest.com/openview/2fc96a6060e1aabede23e8c95fe4b1ad/1?pq-origsite=gscholar&cbl=2032638

Arnedo Cárdenas, A., & Yunes Cañate, K. (2015). Fracking: unconventional oil and gas extraction, and its environmental impact. *Chemical Engineering*. Retrieved from http://hdl.handle.net/10819/2858

Artigas, I. A. (2022, June). The electric car brings out China's weaknesses. *Economic alternatives*(103), 24-25. Retrieved from https://dialnet.unirioja.es/servlet/articulo?codigo=8487335

ARVAL - BNP Paribas Group. (2020). *Practical guide on car emissions and their regulations, global warming, public health, CO2, NOX, WLTP, ...* Retrieved from https://www.arval.es/guia-practica-arval-sobre-las-emisiones-de-automoviles-y-su-normativa-calentamiento-global-salud

Arzata, S. O. (2012). *The world of oil: origin, uses and scenarios.* Fondo de Cultura Económica.

Madrid City Council. (2023). *Electric vehicle charging infrastructure subsidies 2021 - Frequently asked questions.* Retrieved from Madrid City Council: https://www.madrid.es/portales/munimadrid/es/Inicio/Medio-ambiente/Educacion-Ambiental/Subvenciones-infraestructuras-de-recarga-para-vehiculo-electrico-2021-Preguntas-mas-frecuentes/?vgnextoid=71f82c349c288710VgnVCM2000001f4a900aRCRD&vgnextchannel=abd279e

Balcells, D. (2021, 5 12). Toyota develops a revolutionary hydrogen combustion engine. *La Vanguardia*. Retrieved from https://www.lavanguardia.com/motor/vehiculos/coches/20210512/7436245/toyota-coche-motor-combustion-hidrogeno.html

Banagar, I., Mahmoudzadeh Andwari, A., Mehranfar, S., Könnö, J., & Kurvinen, E. (2023). Electric vehicles' powertrain systems architectures design complexity. *Future Technology Open Access Journal, 02*(03), 01-04. Retrieved from https://fupubco.com/futech/article/view/78/57

Barba, J. (2022, November 23). *CREAF.* Retrieved from What is a Climate Shelter?: https://blog.creaf.cat/es/conocimiento/que-es-un-refugio-climatico/

Barría, C. (2020, March 9). Fall in the price of oil: the consequences for Latin America of the fall in the value of crude oil in the midst of the coronavirus crisis. *BBC News World*. Retrieved from https://www.bbc.com/mundo/noticias-51807458

Barría, C. (2021a, November 11). Why car prices have skyrocketed (and what effects it has on the world's economies). *BBC News World*. Retrieved from https://www.bbc.com/mundo/noticias-59151992

BBC News World. (2019, April 1). Which are the countries with the largest oil reserves and why this is not always a sign of wealth. *BBC News World*. Retrieved from https://www.bbc.com/mundo/noticias-47748488

BBC News World. (2022, May 17). Venezuela: The U.S. announces that it is lifting certain sanctions against the Maduro government and allowing dialogue on oil. *BBC News World*. Retrieved from https://www.bbc.com/mundo/noticias-america-latina-61487565

Berenguer Hernández, F. J. (2010). Geopolitics of Energy II. *The New Geopolitics of Energy*, 121-164.

Bermúdez, A. (2021, December 14). Oil: the risks faced by Latin American crude oil producing countries due to the energy transition. *BBC News World*. Retrieved from https://www.bbc.com/mundo/noticias-59589093

Bermúdez, A. (2021, October 15). *Why the price of oil has skyrocketed in the world (and what does the unusual strategy of some producers have to do with it)?* Retrieved from BBC News World: https://www.bbc.com/mundo/noticias-58920072

Biello, D. (2010, August 6). *Where Did the Carter White House's Solar Panels Go?* Retrieved from Scientific American: https://www.scientificamerican.com/article/carter-white-house-solar-panel-array/

Blancafort, R. (2021, May 25). The hydrogen-powered Toyota Corolla completed the 24 Hours of Fuji. *Soymotor*. Retrieved from https://soymotor.com/blogs/rblancafort/el-toyota-corolla-hidrogeno-completo-las-24-horas-de-fuji

Bordino, J. (2021, November 4). *Oil pollution: causes, consequences and how to avoid it*. Retrieved from Green Ecology: https://www.ecologiaverde.com/contaminacion-del-petroleo-causas-consecuencias-y-como-evitarla-3600.html

Boretti, A. (2020, September 3). Hydrogen internal combustion engines to 2030. *International Journal of Hydrogen Energy*, 23692-23703. Retrieved from https://www.sciencedirect.com/science/article/abs/pii/S0360319920321595

Boyd, T. (1968). The Self-Starter. *Technology and Culture, 4*, 585-91. Retrieved from https://doi.org/10.2307/3101900

BP. (2021). *Statistical Review of World Energy 70th edition*. Retrieved from https://www.bp.com/content/dam/bp/business-sites/en/global/corporate/pdfs/energy-economics/statistical-review/bp-stats-review-2021-full-report.pdf

Bravo, E. (2007). Ecological action. *The impacts of oil exploitation on tropical ecosystems and biodiversity., 24(1)*, 35-42. Retrieved from https://d1wqtxts1xzle7.cloudfront.net/46944247/impactos_explotacion_petrolera_esp-libre.pdf?1467429833=&response-content-disposition=inline%3B+filename%3DLOS_IMPACTOS_DE_LA_EXPLOTACION_PETROLERA.pdf&Expires=1683360450&Signature=SYO8WoaBt1sne~lyVpUujSDfOSu

Butter, D. (2015). The war for oil in Syria and Iraq: politics, security and redistribution of resources. *Mediterranean Annual*. Retrieved from https://www.iemed.org/publication/la-guerra-por-el-petroleo-en-siria-y-en-irak-politica-seguridad-y-redistribucion-de-recursos/

Cano, J. L. (2022, December 10). Toyota on the end of the combustion engine: "The electric car should not be the only solution". *Spanish*. Retrieved from https://www.elespanol.com/motor/20221210/toyota-motor-combustion-coche-electrico-no-solucion/724677729_0.html

Cano, V. (2018, 10 11). *What is a hybrid car? Here it is explained in detail*. Retrieved from Autobild: https://www.autobild.es/practicos/que-es-coche-hibrido-313941

CNN. (2020, April 2020). What is OPEC and which countries are part of it? *CNN*. Retrieved from https://cnnespanol.cnn.com/2020/04/10/que-es-la-opep-y-que-paises-la-integran/

CODHES, Consultancy for Human Rights and Displacement. (2013). *The humanitarian crisis in Colombia persists: the disputed Pacific: report of forced displacement in 2012*. Retrieved from https://searchworks.stanford.edu/view/11480726

Council of the European Union. (2023, May 26). *Fit for 55*. Retrieved from https://www.consilium.europa.eu/es/policies/green-deal/fit-for-55-the-eu-plan-for-a-green-transition/

Cooke, P. (2020). Gigafactory logistics in space and time: Tesla's fourth gigafactory and its rivals. *Sustainability, 5*(12), 2044. Retrieved from https://www.mdpi.com/2071-1050/12/5/2044

Corrigan, D. (2022). Electric Vehicle Batteries: Past, Present, and Future. *The Electrochemical Society Interface, 31*(3). Retrieved from https://iopscience.iop.org/article/10.1149/2.F09223IF/meta

Costa, N., Sánchez, L., Anseán, D., & Dubarry, M. (2022, November 25). Li-ion battery degradation modes diagnosis via Convolutional Neural Networks. *Journal of energy storage*. Retrieved from https://www.sciencedirect.com/science/article/pii/S2352152X22015493

Cusumano, M., & Nobeoka, K. (1998). *Thinking beyond lean: how multi-project management is transforming product development at Toyota and other companies*. Simon and Schuster.

dailysportscar.com. (2021, May 23). GTNET Motor Sports Wins Fuji 24 Hours For The Third Time. *dailysportscar.com*. Retrieved from https://www.dailysportscar.com/2021/05/23/gtnet-motor-sports-wins-fuji-24-hours-for-the-third-time.html

Datosmacro.com. (2021). *India – Oil Production*. Retrieved from https://datosmacro.expansion.com/energia-y-medio-ambiente/petroleo/produccion/india

Datosmacro.com. (2021). *Japan - Oil Production*. Retrieved from https://datosmacro.expansion.com/energia-y-medio-ambiente/petroleo/produccion/japon?anio=2021

De Haro, T. (2022, September 2). *The entire readability of the electric car: countries that sell the most, with the most in use and in charging points*. Retrieved from Highway:

https://www.autopista.es/nueva-movilidad/toda-realidad-coche-electrico-paises-mas-venden-en-uso-puntos-carga_263208_102.html

De la Torre, A. (2022, July 12). There are great advances in synthetic fuels for the car. The problem is that no one believes in them. *Xataka*. Retrieved from https://www.xataka.com/movilidad/hay-grandes-avances-combustibles-sinteticos-para-coche-problema-que-nadie-cree-ellos

De la Torre, A. (2023, February 15). "Time will show that our position is the right one": Toyota is very clear about its rejection of the electric car. *Xataka*. Retrieved from https://www.xataka.com/movilidad/toyota-coche-electrico-viven-constante-batalla-estan-dispuestos-a-que-solo-quede-uno

De los Santos Vázquez, S. (1971). *Petroleum Geology*. Retrieved from http://www.ptolomeo.unam.mx:8080/xmlui/handle/132.248.52.100/13266

De Miguel, R., & Planelles, M. (2020, November 18). The United Kingdom brings forward the ban on the sale of petrol and diesel cars to 2030. *El País*. Retrieved from https://elpais.com/clima-y-medio-ambiente/2020-11-18/reino-unido-adelanta-a-2030-la-prohibicion-de-vender-coches-de-gasolina-y-diesel.html

Department of Homeland Security. (2021, February 4). *Global Climate Risk Index 2021*. Retrieved from https://www.dsn.gob.es/en/actualidad/sala-prensa/%C3%ADndice-riesgo-clim%C3%A1tico-global-2021

Devictor, X. (2016, June 22). *The 65 million people displaced by conflict are a challenge for development actors*. Retrieved from https://blogs.worldbank.org/es/voices/65-millones-de-desplazados-por-los-conflictos-un-desafio-para-el-desarrollo

Di Pascoli, S., Femia, A., & Luzzati, T. (2001, August). Natural gas, cars and the environment. A (relatively) 'clean' and cheap fuel looking for users. *Ecological Economics, 38*(2), 179-189. Retrieved from https://www.sciencedirect.com/science/article/abs/pii/S0921800901001744

Di Persio, F., Pérez de Lucia, A., Gramendola, F., Palma, J., Acosta, M., Ferret, R., . . . Escobar, G. (2022). *Recycling of lithium-ion batteries from electric vehicles*. Ministry of Science and Innovation. esoalika technology platform for energy efficiency. Retrieved from https://static.pte-ee.org/media/files/documentacion/itp-01-22-reciclado-de-baterias-de-iones-de-litio-de-vehiculos-electricos-mW.pdf

Díaz, B., & Pareja, R. (2022, August 30). Electric, hybrid, diesel and petrol: how many emissions do they produce in their useful life? *Car and driver*. Retrieved from https://www.caranddriver.com/es/coches/planeta-motor/a30780438/emisiones-contaminantes-segun-tipo-coche/

EFE. (2022, June 14). OPEC acknowledges problems in fulfilling its promise to increase its crude supply. *Five days*. Retrieved from https://cincodias.elpais.com/cincodias/2022/06/14/mercados/1655219863_554644.html

El Periódico. (2023, January 24). What is H2Med? All about its financing and when it will go live. *El Periódico*. Retrieved from https://www.elperiodico.com/es/economia/20230124/h2med-entrara-funcionamiento-2030-81875371

Money. (2019, May 22). The economic impact of autonomous vehicles. *Money*. Retrieved from https://eldinero.com.do/83862/el-impacto-economico-de-los-vehiculos-autonomos/

Electromaps. (2023, February 21). *Charging points*. Retrieved from https://www.electromaps.com/es/puntos-carga

Elvers, B., & Schutze, A. (2021). *Handbook of Fuels: Energy Sources for Transportation 2nd edition*. Wiley. Retrieved from https://www.wiley.com/en-us/Handbook+of+Fuels:+Energy+Sources+for+Transportation,+2nd+Edition-p-9783527333851

Endesa. (2023, March 30). *How do we bring electric mobility closer to all citizens?* Retrieved from Endesa: https://www.endesa.com/es/nuestro-compromiso/transicion-energetica/movilidad-sostenible/acercar-movilidad-electrica-ciudadanos

Endesa. (2023, January 18). *Can electric car batteries be recycled?* Retrieved from The Face e: https://www.endesa.com/es/la-cara-e/economia-circular/reciclaje-baterias-coche-electrico

Enerdata. (2021). Statistics on crude oil production. Retrieved from https://datos.enerdata.net/petroleo-crudo/datos-produccion-energia-mundial.html

Engerer, H., & Horn, M. (2010, February). Natural gas vehicles: An option for Europe. *Energy Policy, 38*(2), 1017-1029. Retrieved from https://www.sciencedirect.com/science/article/abs/pii/S0301421509008131

epdata. (2023, April 5). *Crude oil imports to Spain, data and statistics*. Retrieved from https://www.epdata.es/datos/importaciones-crudo-espana-datos-estadisticas/328

European Automobile Manufacturers' Association. (2022, April 8). *Interactive map - Automobile assembly and production plants in Europe*. Retrieved from https://www.acea.auto/figure/interactive-map-automobile-assembly-and-production-plants-in-europe/

European Environment Agency. (2018). *Electric vehicles from life cycle and circular economy perspectives - TERM 2018: Transport and Environment Reporting Mechanism (TERM) report*. Retrieved from https://www.eea.europa.eu/publications/electric-vehicles-from-life-cycle

Fariza, I. (2022a, March 1). The IEA will release 60 million barrels of crude oil in the face of rising prices caused by the war in Ukraine. *El País*. Retrieved from https://elpais.com/economia/2022-03-01/la-aie-liberara-60-millones-de-barriles-de-crudo-ante-la-crisis-en-ucrania.html

Fariza, I. (2023, June 15). Global oil consumption will set a new high in 2023. *El País*. Retrieved from https://elpais.com/economia/2022-06-15/el-consumo-global-de-petroleo-marcara-un-nuevo-maximo-en-2023.html

Fariza, I. (2023, January 26). Fatih Birol: "Right now there is more uncertainty about diesel than about gas". *El País*. Retrieved from https://elpais.com/economia/2023-01-26/fatih-birol-ahora-mismo-hay-mas-incertidumbre-sobre-el-suministro-de-diesel-que-sobre-el-de-gas-natural.html

Farràs Pérez, L. (2021, May 17). *Electric car batteries, a sustainability problem*. Retrieved from La Vanguardia: https://www.lavanguardia.com/motor/actualidad/20210517/7456307/bateria-coche-electrico-fabricacion-reciclaje-problema-sostenibilidad.html

Fayziyev, P., Ikromov, I., Abduraximov, A., & Dehqonov, Q. (2022, March). Timeline: History of the Electric Car, Trends and the Future Developments. *Eurasian Research Bulletin, 6*, 89-94. Retrieved from https://www.geniusjournals.org/index.php/erb/article/view/888

Fernández Cabezas, F. (2015). Cars powered by hydrogen. *MoleQla: Revista de Ciencias de la Universidad Pablo de Olavide*(17), 110-111. Retrieved from https://www.upo.es/cms1/export/sites/upo/moleqla/documentos/Numero17/Numero_17.pdf

Fueyo, J. (2022). *Blues for a Blue Planet: Civilization's latest challenge to avert the abyss of climate change*. Ediciones B. Retrieved from https://play.google.com/store/books/details?id=lHiHEAAAQBAJ

Gali, J. M. (2013). *Sustainability Marketing*. Profit Editorial.

García Ropero, J. (2023, January 22). Germany joins the green hydrogen interconnection of Spain and France. *El País*. Retrieved from https://cincodias.elpais.com/cincodias/2023/01/22/economia/1674409485_266723.html

García, G. (2020, June 22). *Reusing electric car batteries is better than recycling them, albeit with nuances*. Recovered from Hybrids and electrics: https://www.hibridosyelectricos.com/coches/reutilizar-baterias-coches-electricos-reciclar_36074_102.html

García, G. (2022, February 16). Euro 7: Europe is leading the way to ban combustion cars by 2025. *Hybrid and electric*. Retrieved from https://www.hibridosyelectricos.com/coches/euro-7-europa-prohibicion-coches-combustion-2025_40100_102.html

García, P., & Ramírez, R. (2008). The International Energy Agency's Long-Term Energy Outlook: World Energy Outlook 2008. *ICE Economic Bulletin, 2955*. Retrieved from https://www.researchgate.net/profile/Antonio-Merino-5/publication/28237742_Perspectivas_energeticas_a_largo_plazo_de_la_Agencia_Internacional_de_la_Energia_World_Energy_Outlook_2008/links/0a85e53c3da3c5c655000000/Perspectivas-energeticas-a-largo-plazo-de-

García-Astillero, A. (2019, April 24). *Environmental impact of oil and natural gas*. Retrieved from https://www.ecologiaverde.com/impacto-ambiental-del-petroleo-y-el-gas-natural-1658.html

Garmendia, X. (2016). Atoms against volts and bits. The end of the era of oil, renewable energies and the new energy model. *Grand Place*, 62. Retrieved from https://marioonaindiafundazioa.org/wp-content/uploads/2021/06/GRAND-PLACE-5.pdf

Germanwatch. (2023). *Global Climate Risk Index*. Retrieved from https://www.germanwatch.org/en/cri

Government of Spain. (2022, December 16). *H2Med application submitted as an EU Project of Common Interest*. Retrieved from https://www.lamoncloa.gob.es/lang/en/gobierno/news/Paginas/2022/20221216_h2med -green-hydrogen-hub.aspx

Gonzales, G., Ordolez, C., & Vásquez-Velásquez, C. (2022, 03 07). Oil spill in Ventanilla, Callao, January 2022. *Journal of the Peruvian Society of Internal Medicine, 35*(1), 47-49. Retrieved from https://revistamedicinainterna.net/index.php/spmi/article/view/658

González Pérez, R. (2022). *Hydrogen. Fuel cell: Gases in the air*. Ediciones Díaz de Santos.

González Rodríguez, L., Blanco Trejo, F., & González Hervias, R. (2022). Mental health and political migration Latin America. Individual experiences and coping styles. *NURE research: Scientific Journal of Nursing*(119). Retrieved from https://dialnet.unirioja.es/servlet/articulo?codigo=8584895

González Trujillo, A., & Cabrera Rodríguez, D. (2019). *The U.S. Intervention in the Middle East: The Case of Iraq (2003)*. Retrieved from https://riull.ull.es/xmlui/bitstream/handle/915/12263/La%20intervencion%20de%20Est ados%20Unidos%20en%20Oriente%20Medio%20El%20caso%20de%20Irak%20(2003).pdf?sequence=1

Gonzalez, K. (2013). *From the EV1 to the Chevy Volt: A re-electrification of the American automobile industry*. University of Southern California.

Granda, M. (2022, February 21). This is the distribution of charging points for electric cars in Spain. *Five days*. Retrieved from https://cincodias.elpais.com/cincodias/2022/02/18/companias/1645202903_239316.ht ml

Greenpeace. (2012). *Environmental impacts of petroleum*. Retrieved from https://www.greenpeace.org/static/planet4-mexico-stateless/2018/11/cd1362c6-cd1362c6-impactos_ambientales_petroleo.pdf

Greenpeace. (2020). *By 2025, batteries retired from China's new energy vehicles can be used to cover the backup power demand of 5G base stations across the country*. Retrieved from https://www.greenpeace.org.cn/2020/10/29/ev-battery-report-20201029/

Guijarro, J. R. (2022). The return of geopolitical risk: economic effects of the war in Ukraine. *Cuadernos de Información Económica., 288*, 1-10.

Gutiérrez, I. (2019, July 5). What is OPEC and what is its influence on the economy? *Very financial*. Retrieved from https://muyfinanciero.com/organizaciones/gobierno/opep/

Guzmán, R. (2016). Hydraulic fracturing and the Monterrey VI hydraulic project; irreversible damage to the environment. *Plurality and consensus*. Retrieved from http://www.revista.ibd.senado.gob.mx/index.php/PluralidadyConsenso/article/view/333/339

Haidai, O. R. (2022). Mine Field Preparation and Coal Mining in Western Donbas: Energy Security of Ukraine - A case Study. *Energies, 15(13)*. doi:10.3390/en15134653

Heredia Martínez, J. (2022, 11 24). Universitat Politècnica de València. *Design of the electrical system of a grid-connected charging station with photovoltaic solar energy input (Doctoral Dissertation)*. Retrieved from https://riunet.upv.es/handle/10251/190128

Hybrid and Electric. (2020, April 15). *Keys to developing a global charging infrastructure for electric cars*. Retrieved from https://www.hibridosyelectricos.com/coches/claves-desarrollo-infraestructura-recarga-global-vehiculos-electricos_33799_102.html

Hontoria, N. (2020, September 30). *Electric car batteries: all the questions about their duration, manufacture and recycling when they are no longer useful*. Retrieved from Uppers: https://www.uppers.es/motor-y-movilidad/motor/baterias-coches-electricos-reciclaje-reutilizacion_18_3018495279.html

Horowitz, J. (2021, October 13). Peak demand for oil is coming, but that won't save the world. *CNN in Spanish*. Retrieved from https://cnnespanol.cnn.com/2021/10/13/el-pico-de-demanda-del-petroleo-se-acerca-pero-eso-no-salvara-al-mundo/

Hosteltur. (2022, August 30). *Travelling by electric car in Europe: the best and worst prepared countries*. Retrieved from Hosteltur: https://www.hosteltur.com/152735_viajar-en-coche-electrico-por-europa-los-paises-mejor-y-peor-preparados.html

Hyundai. (2021, January 8). *Regenerative braking, what is it and how to take advantage of it?* Retrieved from Zona Eco by Hyundai: https://www.hyundai.com/es/zonaeco/eco-drive/tecnologia/frenada-regenerativa

Iberdrola. (2021). *Electric vehicle charging stations*. Retrieved from https://www.iberdrola.com/sostenibilidad/estaciones-carga-vehiculos-electricos

Iberdrola. (2023, May 4). *Battery recycling, a new life outside the car*. Retrieved from Iberdrola: https://www.iberdrola.es/blog/transporte/vida-util-bateria-coche-electrico

Iglesias, G. F. (2015). *History of taxis and LPG fuel in Spain*. Ediciones de Buena Tinta.

Impemba, M., & Gazzera, M. (2019). Tourism and Fracking. Journey to the heart of Vaca Muerta: Who is left with the impacts? *Revista Gestión I+ D*, 110-132.

Ingebor. (2017). Electric car incentives are growing across Europe. *forocoqueselectricos*. Retrieved from https://forocoqueselectricos.com/2017/09/incentivos-al-coche-electrico-europa.html

International Energy Agency. (2022). *World Energy Outlook 2022*. Paris. Retrieved from https://www.iea.org/reports/world-energy-outlook-2022/executive-summary?language=es

Jiménez, F. (2008). Safety in the distribution and handling of LPG. (B. v. environmental, Compiler) Retrieved from https://sistemamid.com.ar/panel/uploads/biblioteca/2016-06-06_08-53-57135110.pdf

Jiménez, J. C. (2019). The time for green hydrogen. *Current gas*(153), 24-30. Retrieved from https://www.gasrenovable.org/docs/hidrogeno_renovable/Reportaje-La_hora_del_Hidrogeno_verde.pdf

Jingyuan, Z., & Burke, A. (2022, September 24). Electric Vehicle Batteries: Status and Perspectives of Data-Driven Diagnosis and Prognosis. *Batteries*. Retrieved from https://www.mdpi.com/2313-0105/8/10/142

Katouzian, H. (2022). The Anglo-Iranian oil crisis revisited: Iran's rejection of the World Bank intervention and the 1953 coup. *British Journal of Middle Eastern Studies*, 1-10. Retrieved from https://www.tandfonline.com/doi/full/10.1080/13530194.2022.2128046

Keith, D., & Farrell, A. (2003, July 18). Rethinking Hydrogen Cars. *Science, 301*(5631), 315-316. Retrieved from https://www.science.org/doi/full/10.1126/science.1084294

Khodunov, A. (2022). The Orange Revolution in Ukraine. Springer International Publishing.

Klare, M. (2020). *Wars for resources: the future scenario of global conflict* (Vol. 21). LAVP Editions.

Klein, B. (2021, November 23). Biden announces the release of barrels from the Strategic Petroleum Reserve, but warns that prices will not come down immediately. *CNN in Spanish*. Retrieved from https://cnnespanol.cnn.com/2021/11/23/liberacion-50-millones-barriles-petroleo-reserva-estrategica-trax/

Kumar Singla, M., Nijhawan, P., & Singh Oberoi, A. (2021). Hydrogen fuel and fuel cell technology for cleaner future: a review. *Environmental Science and Pollution Research*. Retrieved from https://link.springer.com/article/10.1007/s11356-020-12231-8

Labandeira, X. (2012). Economics for Energy Working Paper. *Energy system and climate change: technological and regulatory prospective.*

Lapatilla. (2022, August 10). This is the oil market for Venezuela, according to economist Rafael Quiroz. *Lapatilla*. Retrieved from https://www.lapatilla.com/2022/08/10/asi-esta-el-mercado-petrolero-para-venezuela/

Laurens Tamon, M. (2023). Gulf War (1980-1991): Causality and Chronological Study Incident. *International Journal of Multicultural and Multireligious Understanding, 10*(1), 129-141. Retrieved from https://ijmmu.com/index.php/ijmmu/article/view/4258

Liadze, I., Macchiarelli, C., Mortimer-Lee, P., & Sanchez Juanino, P. (2022, September 14). Economic costs of the Russia-Ukraine war. *The World Economy, 46*(4), 874-886. Retrieved from https://onlinelibrary.wiley.com/doi/abs/10.1111/twec.13336

Link, A., O'Connor, A., Scott, T., Casey, S., Loomis, R., & Lynn David, J. (2013). *Benefit-Cost Evaluation of U.S. DOE Investment in Energy Storage Technologies fro Hybrid and Electric Cars and Trucks*. U.S. Department of Energy, Office of Energy Efficiency and Renewable Energy. Retrieved from https://www.energy.gov/eere/analysis/articles/benefit-cost-evaluation-us-doe-investment-energy-storage-technologies-hybrid

Lohse-Busch, H., Stutenberg, K., Duoba, M., & Iliev, S. (2018). *Technology Assessment Of A Fuel Cell Vehicle: 2017 Toyota Mirai*. Office of Scientific and Technical Information, U.S. Department of Energy, Argonne, IL. Retrieved from https://www.osti.gov/biblio/1463251

López Redondo, N. (2020, February 26). How much electric car batteries really degrade. *movilidadelectrica.com*. Retrieved from https://movilidadelectrica.com/como-se-degradan-baterias-coches-electricos-54004-2

López, N. (2023, April 11). This is how long electric car batteries really last. *Autobild.es*. Retrieved from https://www.autobild.es/noticias/dura-verdad-baterias-coches-electricos-1228390

Lóppez, N. (2022, June 23). *How a hydrogen car (or fuel cell) works*. Retrieved from Autobild: https://www.autobild.es/practicos/como-funciona-coche-hidrogeno-pila-combustible-1082221

LugEnergy. (2022). *Electric vehicle charging modes*. Retrieved from https://www.lugenergy.com/modos-de-recarga-vehiculos-electricos/

Mader, J. (2006). *Battery Powered vehicles: don't rule them out*. University of Michigan, Transportation Energy Center (TEC). Retrieved from https://cheresearch.engin.umich.edu/schwank/people/Papers/BatteryPower.pdf

Maldita.es. (2023, April 25). *What are climate shelters and how do they help vulnerable people protect themselves from the heat?* Retrieved from https://maldita.es/malditaciencia/20230425/que-son-refugios-climaticos/

Maldonado, Y. (2021, February 17). *10 advantages and disadvantages of oil*. Retrieved from Geologiaweb: https://geologiaweb.com/recursos-naturales/ventajas-desventajas-petroleo/

Mandev, A., Plötz, P., Sprei, F., & Tal, G. (2022, September 1). Empirical charging behavior of plug-in hybrid electric vehicles. *Applied Energy, 321*. Retrieved from https://www.sciencedirect.com/science/article/pii/S0306261922006481

Manos Unidas. (n.d.). *Climate Refugees*. Retrieved from https://www.manosunidas.org/observatorio/cambio-climatico/refugiados-climaticos

Mapfre. (2022, July 7). *Second life of electric car batteries*. Retrieved from Motor Blogs Mapfre: https://www.motor.mapfre.es/coches/noticias-coches/baterias-coches-electricos/

Mapfre. (2023, February 27). *Legislation for electric cars in Europe*. Retrieved from Mapfre: https://www.mapfre.es/particulares/seguros-de-coche/articulos/legislacion-coches-electricos-europa/

Martín Spuch, J. M. (2021, October 4). Battery degradation: the biggest fear of the electric car? *I am Motor*. Retrieved from https://soymotor.com/coches/articulos/degradacion-de-la-bateria-coche-electrico-990993

Martínez Rull, E. (2022, March 25). Europe is burning coal again. *La Razón*. Retrieved from https://www.larazon.es/medio-ambiente/20220325/mdglzzil2vctxclay7fpzaesyu.html

Martins, H., Henriques, C., Figueira, J., Silva, C., & Costa, A. (2022, June). Assessing policy interventions to stimulate the transition of electric vehicle technology in the European Union. *Socio-Economic Planning Sciences, 87*. Retrieved from https://www.sciencedirect.com/science/article/pii/S0038012122003123

Ministry of Transport, Mobility and Urban Agenda. (2022, June 14). *The Government approves a regulation that requires a minimum infrastructure in buildings for the recharging of electric vehicles*. Retrieved from https://www.mitma.gob.es/el-ministerio/sala-de-prensa/noticias/mar-14062022-1713

Ministry for the ecological transition and the demographic challenge. (2020, June 22). *Climate change: Transport sector*. Retrieved from https://www.miteco.gob.es/es/cambio-climatico/temas/mitigacion-politicas-y-medidas/transporte.aspx

Ministry for the ecological transition and the demographic challenge. (2020). *FUGITIVE EMISSIONS FROM THE EXPLORATION, EXTRACTION, FIRST TREATMENT AND LOADING OF LIQUID FOSSIL FUELS IN ONSHORE FACILITIES*. Retrieved from https://www.miteco.gob.es/es/calidad-y-evaluacion-ambiental/temas/sistema-espanol-de-inventario-sei-/050201-fugitivas-extracc-combust-liq-tierra_tcm30-430157.pdf

Ministry for the Ecological Transition and the Demographic Challenge. (2021). *MOVES III Programme*. Retrieved from https://www.idae.es/ayudas-y-financiacion/para-movilidad-y-vehiculos/programa-moves-iii

Mokesioliwa, F., Mladen, B., & Sairabh, S. (2022). A Review of the Impact of Battery Degradation on Energy Management Systems with a Special Emphasis on Electric Vehicles. *Energies*. Retrieved from https://www.mdpi.com/1996-1073/15/16/5889

Montenegro, V., & Torra, S. (2015, December 13). Dissertation: The Automobile Industry and the Environment. Retrieved from https://comercioexterior.ub.edu/tesina/tesinasaprobadas/1415/TesinaMontenegroVictor.pdf

Montes, L. (2019, December 28). More than 200 million people could become climate refugees in 2022 – these are the regions that will be hit the hardest. *Business Insider*. Retrieved from https://www.businessinsider.es/son-refugiados-climaticos-cuales-son-regiones-afectadas-550729

Moscoso, B. (2021, May 27). The Toyota Corolla H2 concept completes the 24 Hours of Fuji. *Octane*. Retrieved from https://octanosmedia.com/2021/05/el-toyota-corolla-h2-concept-completa-las-24-horas-de-fuji-de-la-super-taikyu-series/

Murias, D. (2021, December 3). The hydrogen Toyota GR Yaris H2 is the last and fine hope of the internal combustion engine. *Motorpassion*. Retrieved from https://www.motorpasion.com/toyota/toyota-gr-yaris-h2-hidrogeno-ultima-fina-esperanza-motor-combustion-interna

United Nations. (2021). *Migration*. Retrieved from https://www.un.org/es/global-issues/migration

Næss, H., & Tjønndal, A. (2021). *Innovation, Sustainability and Management in Motorsports: The Case of Formula E.* Palgrave Macmillan. Retrieved from https://library.oapen.org/handle/20.500.12657/49508

Nagle, P. &. (2021, November 2). *Oil market developments: rising prices amid broader energy price rises.* Retrieved from World Bank Blogs: https://blogs.worldbank.org/es/datos/evolucion-del-mercado-del-petroleo-y-los-precios-de-la-energia

Nakazawa, S. (2011). The Nissan LEAF electric powertrain. *32. International Vienna motor symposium*, (pp. 330-341). Vienna. Retrieved from https://www.osti.gov/etdeweb/biblio/21478310

National Geographic. (2022). What is the environmental impact of cars? *National Geographic.* Retrieved from https://www.nationalgeographic.es/medio-ambiente/que-impacto-medioambiental-tienen-los-coches

Okamura, M., & Takaoka, T. (2022, November 12). The Evolution of Electric Components in Prius. *IEEJ Journal of Industry Applications, 1*(11), 1-6. Retrieved from https://www.jstage.jst.go.jp/article/ieejjia/11/1/11_21007126/_article/-char/ja/

OPEC. (n.d.). *OPEC Mission.* Retrieved from OPEC: https://www.opec.org/opec_web/en/about_us/24.htm

Orhan, E. (2022). The Effects of the Russia - Ukraine War on Global Trade. *Journal of International Trade, Logistics and Law*, 141-146.

Overland, I. (2016, April). Energy: The missing link in globalization. *Energy Research & Social Science, 14*, 122-130. Retrieved from https://www.sciencedirect.com/science/article/pii/S2214629616300093?via%3Dihub

Paredes, N. (2021, March 22). Crisis in Venezuela | How long it would take the country to revive its oil industry (and why it is vital for economic recovery). *BBC News World.* Retrieved from https://www.bbc.com/mundo/noticias-america-latina-56351173

European Parliament. (2022, December). *Environmental policy: general principles and basic framework.* Retrieved from Thematic Factsheets on the European Union: https://www.europarl.europa.eu/factsheets/es/sheet/71/la-politica-de-medio-ambiente-principios-generales-y-marco-basico

European Parliament News. (2022, October 28). *The ban on the sale of new petrol and diesel cars from 2035 in the EU.* Retrieved from European Parliament News: https://www.europarl.europa.eu/news/es/headlines/economy/20221019STO44572/la-prohibicion-de-vender-nuevos-coches-de-gasolina-y-diesel-a-partir-de-2035

Passsarello, H., & Radio France Internationale. (2021, November 1). *COP26: The challenge of recycling lithium batteries in electric cars.* Retrieved from Radio France Internationale: https://www.rfi.fr/es/programas/enfoque-internacional/20211101-cop26-el-desaf%C3%ADo-de-reciclar-las-bater%C3%ADas-de-litio-de-los-coches-el%C3%A9ctricos

Pérez, C., & Peraza, G. (2022). *Oil, power and civilization.* RUTH.

Pérez, R. (2022a, June 9). The end of combustion cars: what will happen in 2035? *El País Motor*. Retrieved from https://motor.elpais.com/actualidad/el-fin-de-los-coches-de-combustion-que-pasara-en-2035/

Pérez, R. (2023, February 9). Why Toyota rejects the electric car. *El País*. Retrieved from https://motor.elpais.com/actualidad/la-marca-superventas-que-no-cree-en-el-coche-electrico/

Pérez, R. (2023a, February 15). Will diesel and petrol cars be able to be used after 2035? *El País Motor*. Retrieved from https://motor.elpais.com/actualidad/podran-conducirse-los-coches-de-combustion-despues-de-2035/

Phillips, C. (2019, November 1). The Battle for Syria: International Rivalry in the New Middle East – a reply. *Global Discourse*, 759-766. Retrieved from https://bristoluniversitypressdigital.com/view/journals/gd/9/4/article-p759.xml?rskey=bDDVZO&result=127&utm_source=TrendMD&utm_medium=cpc&utm_campaign=Global_Discourse%253A_An_interdisciplinary_journal_of_current_affairs_TrendMD_0

Plaza, D. (2022, March 29). These are all the car factories in Spain. *Motor.es*. Retrieved from https://www.motor.es/noticias/fabricas-coches-espana-202285950.html

Plaza, D. (n.d.). *The electric car: definition and types*. Retrieved from motor.es: https://www.motor.es/que-es/coche-electrico

Portaluppi, A. (2021, December 31). *Country by country, the 2021 numbers in charging infrastructure for electric vehicles in Latin America*. Retrieved from https://portalmovilidad.com/pais-por-pais-los-numeros-del-2021-en-infraestructura-de-carga-para-vehiculos-electricos-en-latinoamerica/

Prego, C. (2022, February 20). Toyota and Yamaha are developing a hydrogen V8 engine. Its objective: to save internal combustion. *Xataka*. Retrieved from https://www.xataka.com/vehiculos/toyota-yamaha-estan-desarrollando-motor-v8-hidrogeno-su-objetivo-salvar-combustion-interna

PwC Price Waterhouse Coopers. (2022). *eReadiness 2022*. Retrieved from https://www.pwc.es/es/publicaciones/automocion/assets/strategyand-ereadiness-study-2022.pdf

RACE. (2023). *How do you recharge an electric car?* Retrieved from RACE: https://www.race.es/informacion-para-cargar-coche-electrico

Rafael, G. J. (2018, September 10). *Oil: its geopolitical importance*. Retrieved from https://www.linkedin.com/pulse/el-petr%C3%B3leo-su-importancia-geopol%C3%ADtica-jos%C3%A9-rafael-gamero-lanz/

ROYAL SPANISH ACADEMY. (2023, 06 06). *Dictionary of the Spanish Language, 23rd ed. [version 23.6 online]*. Retrieved from https://dle.rae.es

National Geographic Newsroom. (2018, January 4). *Is it more polluting to manufacture an electric car than a conventional one?* Retrieved from https://www.nationalgeographic.es/coche-electrico-contaminacion

Rejón, R. (2018, November 19). Banning the sale of diesel and petrol cars by 2040 is not revolutionary, but an international trend. *eldiario.es*. Retrieved from https://www.eldiario.es/sociedad/prohibicion-diesel-worldwide_1_1830294.html

Repsol YPF. (n.d.). *How many Autogas or LPG petrol stations are there in Spain?* Retrieved from https://www.repsol.es/particulares/faqs/carburantes/cuantas-gasolineras-de-autogas-o-glp-hay-en-espana/

REUTERS. (2023, April 2). v. *El País*. Retrieved from https://elpais.com/economia/2023-04-02/la-opep-anuncia-por-sorpresa-un-recorte-de-11-millones-de-barriles-diarios-hasta-2024.html

Reynoso, J. (2023). *Environmental Protection of the Air*. UNAM, Faculty of Chemistry. Retrieved from https://play.google.com/store/books/details?id=YXyyEAAAQBAJ

Romero Places, M., Diego Olite, F., & Álvarez Toste, M. (2006). Air pollution: its impact as a health problem. *Cuban Journal of Hygiene and Epidemiology*. Retrieved from http://scielo.sld.cu/scielo.php?script=sci_arttext&pid=s1561-30032006000200008

Romero, F. (2021, April 23). Toyota develops a hydrogen-fueled combustion engine. *motor.es*. Retrieved from https://www.motor.es/noticias/toyota-motor-combustion-hidrogeno-202177395.html

RTVE.es/AGENCIAS. (2023, February 3). *Spain imported more crude oil but reduced its dependence on Russia by 73% in 2022*. Retrieved from https://www.rtve.es/noticias/20230203/espana-importo-mas-crudo-pero-redujo-su-dependencia-rusia-73-2022/2420767.shtml

Rueda, A. (2022, September 4). *Can electric car batteries be recycled?* Retrieved from El motor, El País: https://motor.elpais.com/tecnologia/se-pueden-reciclar-las-baterias-de-los-coches-electricos/

Ruiz Ruiz, N., & Santana Rivas, L. (2020). *The relationship between forced displacement, mining and the geographies of accumulation by dispossession in the last 20 years in Colombia*. Retrieved from https://files.alapop.org/congreso6/files/pdf/alap_2014_final349.pdf

Ruiz, N. (2020). Mortality due to malnutrition in children and its relationship with territorial determinants (Colombia, 2003-2012). *CIST2020*. Retrieved from https://hal.science/hal-03115024/

Sadeghian, O., Oshnoei, A., Mohammadi-ivatloo, B., Vahidinasab, V., & Anvari-Moghaddam, A. (2022, October). A comprehensive review on electric vehicles smart charging: Solutions, strategies, technologies, and challenges. *Journal of Energy Storage, 54*. Retrieved from https://www.sciencedirect.com/science/article/abs/pii/S2352152X22012403

Santiago, A., Carrillo, B., & Martella, B. (1984). *Petroleum geology of Mexico*. Retrieved from https://d1wqtxts1xzle7.cloudfront.net/34981687/GEOLOGIA_PETROLERA_DE_MEXICO-libre.pdf?1412333639=&response-content-disposition=inline%3B+filename%3DGEOLOGIA_PETROLERA_DE_MEXICO.pdf&Expires=1683359946&Signature=RgysovAmDMpUbaJt5F5yD16Z~xGbLJu00PvWA3MWxWOY1

Sanz, J. C. (2019, December 10). Oil by territory in Syria. *El País*. Retrieved from https://elpais.com/internacional/2019/12/09/actualidad/1575919898_140549.html

Segoviano, D. (2016, June 20). OPEC: divided, but influential. *Millennium*. Retrieved from https://www.milenio.com/negocios/la-opep-dividida-pero-influyente

Sopeña, J. E. (2001). *LPG-Instruction Manual*. CEPSA ELF GAS.

Spadaro, P. A. (2020, May 10). Climate Change, Environmental Terrorism, Eco-Terrorism and Emerging Threats. *Journal of Strategic Security, 13*(4), 58-80. Retrieved from https://www.jstor.org/stable/26965518

Statista. (2022). *Ranking of countries with the highest estimated gross domestic product (GDP) from 2021 to 2027*. Retrieved from https://es.statista.com/estadisticas/600234/ranking-de-paises-con-el-producto-interior-bruto-pib-mas-alto-en/

Statista. (2023). *Ranking of the countries that recorded the highest oil consumption in 2021*. Retrieved from https://es.statista.com/estadisticas/601254/consumo-de-petroleo-por-paises/

Statista Research Department. (2023). *Market share of major oil-consuming countries 2010-2021*. Retrieved from https://es.statista.com/estadisticas/600906/principales-paises-consumidores-de-petroleo-a-nivel-mundial/

Statista Research Department. (2023). *Oil reserves in selected countries from 1990 to 2020*. Retrieved from https://es.statista.com/estadisticas/635340/reservas-de-petroleo-por-paises/

Teat Arquitectos. (2023, May 3). *What are climate shelters?* Retrieved from https://www.teat.es/noticias/que-son-los-refugios-climaticos/

Tomé, C. (2019, May 2). The oil peak. *Cuaderno de Cultura Científica*. Retrieved from https://culturacientifica.com/2019/05/03/el-pico-petrolero/

Torres Benayas, V. (2022, August 5). What is a climate shelter and where to find them in Spain? *El País*. Retrieved from https://elpais.com/clima-y-medio-ambiente/2022-08-05/que-es-un-refugio-climatico-y-donde-encontrarlos-en-espana.html

Toyota Spain. (2021, December 14). *Toyota introduces a full range of battery electric vehicles globally*. Retrieved from https://prensa.toyota.es/toyota-presenta-una-gama-completa-de-vehiculos-electricos-de-bateria-a-nivel-mundial/

Toyota Spain Press. (2020, December 3). *New Toyota Mirai: this is how the second generation of Toyota's fuel cell sedan is presented*. Retrieved from Toyota Spain Press Room: https://prensa.toyota.es/nuevo-toyota-mirai-asi-se-presenta-la-segunda-generacion-de-la-berlina-de-pila-de-combustible-de-toyota/

Toyota Spain. (n.d.). *Hybrid and electric vehicles*. Retrieved from https://www.toyota.es/electrificacion

Toyota Motor Corporation. (2021, December 14). Media Briefing on Battery EV Strategies. Tokyo, Japan. Retrieved from https://www.youtube.com/watch?v=M3C07uxyaWQ

Ubilla, C., & Yohannessen, K. (2017). Air pollution: effects on respiratory health in children. *Revista Médica Clínica Las Condes*. Retrieved from https://www.sciencedirect.com/science/article/pii/S0716864017300214

UN - United Nations. (2023). *Causes and effects of climate change*. Retrieved from https://www.un.org/es/climatechange/science/causes-effects-climate-change

UNHCR UNHCR. (2016). *Global Trends: Forced Displacement in 2015. Forced to flee*. Retrieved from https://www.acnur.org/fileadmin/Documentos/Publicaciones/2016/10627.pdf

UNHCR UNHCR. (2019, June 28). *Climate change and disasters are causing more and more displacement*. Retrieved from https://eacnur.org/es/actualidad/noticias/emergencias/refugiados-climaticos

UNHCR UNHCR. (2021, April 19). *World map of climate displacement*. Retrieved from https://eacnur.org/es/actualidad/noticias/mapa-mundial-desplazamiento-emergencia-climatica

UNHCR UNHCR. (2021, January 6). *Countries and territories with the most internal displacement due to natural disasters in 2020*. Retrieved from https://eacnur.org/es/actualidad/noticias/emergencias/paises-territorios-desplazamientos-clima-desastres-naturales-2020

Valencia, J. (2020). *Climate Shelters: concept, background and need to include them in the Draft Framework Law on Climate Change*. Parliamentary Minutes for the discussion of the Framework Law in the first Constitutional procedure, Austral Patagonia Program of the Universidad Austral de Chile, Valdivia, Chile. Retrieved from https://obtienearchivo.bcn.cl/obtienearchivo?id=documentos/10221.1/80487/1/Minuta_LEY-MARCO-DE-CAMBIO-CLIMATICO.pdf

Vargas, J. G. (2017). Impact of the price of oil on the GDP of the countries of the Pacific Alliance. *Revista Financias y Política Económica*, https://redalyc.org/journal/3235/323553607003/html/.

Vega, L. (2023, February 2). Toyota uses science to argue its rejection of the electric car. *Business Insider*. Retrieved from https://www.businessinsider.es/toyota-utiliza-ciencia-argumentar-rechazo-coche-electrico-1194240

Verhelst, S., Maesschalck, P., Rombaut, N., & Sierens, R. (2009, May). Increasing the power output of hydrogen internal combustion engines by means of supercharging and exhaust gas recirculation. *International Journal of Hydrogen Energy, 34*(10), 4406-4412. Retrieved from https://www.sciencedirect.com/science/article/abs/pii/S036031990900442X

Viñuela, S. (2019, November 12). *What happens to electric car batteries when they run out?* Retrieved from Autobild.es: https://www.autobild.es/reportajes/pasa-baterias-coches-electricos-cuando-agotan-527121

Volkswagen. (2023). *Electric car vs petrol: Cost comparison*. Retrieved from https://www.volkswagen.es/es/revista/modelos/coche-electrico-vs-gasolina.html

Wang, H., He, H., Yungfei, B., & Hongwei, Y. (2022, August 15). Parameterized deep Q-network based energy management with balanced energy economy and battery life for hybrid electric vehicles. *Applied energy.* Retrieved from https://www.sciencedirect.com/science/article/abs/pii/S0306261922006274

Wauquier, J. (2004). *Oil refining: crude oil, petroleum products, manufacturing schemes.* Ediciones Díaz de Santos.

White, I. C. (1885). The geology of natural gas. *Science,* 521. Retrieved from https://www.science.org/doi/10.1126/science.ns-5.125.521

Winckler, J., & Del Castillo Pantoja, G. (2020). Anthropogenic climate change and forced displacement. *Politics and Strategy Magazine.* Retrieved from https://dialnet.unirioja.es/servlet/articulo?codigo=7708747

Wolff, B., Schlagwein, D., & Schoder, D. (2022). How Resource Openness Supports Industrial Transformation and Competitive Advantage: The Revelatory Case of Tesla. *Pacific Asia Conference on Information Systems,* (p. 1). Retrieved from https://www.researchgate.net/profile/Bastian-Wolff/publication/361228411_How_Resource_Openness_Supports_Industrial_Transformation_and_Competitive_Advantage_The_Revelatory_Case_of_Tesla/links/62a48002416ec50bdb1c5983/How-Resource-Openness-Supports-Industri

Wu, Y., Han, G., & Xie, W. (2022, February 25). Economics of Battery Swapping for Electric Vehicles—Simulation-Based Analysis. *Energies, 15*(5), 1714. Retrieved from https://www.mdpi.com/1996-1073/15/5/1714

Yergin, D. (1992). *History of oil.* Plaza y Janés editores, S.A.

YPF. (2022). *Oil extraction methods.* Retrieved from https://www.ypf.com/energiaypf/Metodosdeextraccion/extraccion_convencional.html